JOURNAL OF SCIENTIFIC EXPLORATION
A Publication of the [barcode] ation

(ISSN 0892-3310) published qua[r] S0-BYN-381

Editorial Office: Journal@ScientificExploration.org

Manuscript Submission: http://journalofscientificexploration.org/index.php/jse/

Editor-in-Chief: Stephen E. Braude, University of Maryland Baltimore County

Managing Editor: Kathleen E. Erickson, San Jose State University, California
Assistant Managing Editor: Elissa Hoeger, Princeton, NJ

Associate Editors
Carlos S. Alvarado, Parapsychology Foundation, New York, New York
Imants Barušs, University of Western Ontario, London, Ontario, Canada
Daryl Bem, Ph.D., Cornell University, Ithaca, New York
Robert Bobrow, Stony Brook University, Stony Brook, New York
Jeremy Drake, Harvard–Smithsonian Center for Astrophysics, Cambridge, Massachusetts
Michael Ibison, Institute for Advanced Studies, Austin, Texas
Roger D. Nelson, Princeton University, Princeton, New Jersey
Mark Rodeghier, Center for UFO Studies, Chicago, Illinois
Harald Walach, Viadrina European University, Frankfurt, Germany

Publications Committee Chair: Garret Moddel, University of Colorado Boulder

SUBSCRIPTIONS & PREVIOUS JOURNAL ISSUES: Order forms on back pages or at scientific-exploration.org.

Society for Scientific Exploration—https://www.scientificexploration.org

Journal of Scientific Exploration (ISSN 0892-3310), an open access, peer-reviewed journal, is published quarterly in March, June, September, and December by the Society for Scientific Exploration, P. O. Box 8012 Princeton, NJ 08543 USA. The Journal is free to everyone. Society Members may purchase print subscriptions for $60 per year. Library print subscriptions are $165 per year.

JOURNAL OF SCIENTIFIC EXPLORATION
A Publication of the Society for Scientific Exploration

AIMS AND SCOPE: The *Journal of Scientific Exploration* is an Open Access journal, which publishes material consistent with the Society's mission: to provide a professional forum for critical discussion of topics that are for various reasons ignored or studied inadequately within mainstream science, and to promote improved understanding of social and intellectual factors that limit the scope of scientific inquiry. Topics of interest cover a wide spectrum, ranging from apparent anomalies in well-established disciplines to rogue phenomena that seem to belong to no established discipline, as well as philosophical issues about the connections among disciplines. The *Journal* publishes research articles, review articles, essays, commentaries, guest editorials, historical perspectives, obituaries, book reviews, and letters or commentaries pertaining to previously published material.

The Journal of Scientific Exploration is indexed in Scopus, Elsevier Abstracts, and the Directory of Open Access Journals (DOAJ).

https://doi.org/10.31275/2020.1867 for this whole issue PDF, JSE 34:2, Summer 2020.

JOURNAL OF SCIENTIFIC EXPLORATION
A Publication of the Society for Scientific Exploration

Volume 34, Number 2   2020

EDITORIAL

RESEARCH ARTICLES

COMMENTARY

BOOK REVIEWS

SSE NEWS

Journal of Scientific Exploration, Vol. 34, No. 2, pp. 199–208, 2020   0892-3310/20

EDITORIAL

Does Telepathy Threaten Mental Privacy?

Stephen E. Braude

https://doi.org/10.31275/2020/1829

A long-standing concern (or at least a belief) about ESP, held by both skeptics and believers in the paranormal, is that if telepathy really occurs, then it might pose a threat to mental privacy. And it's easy enough to see what motivates that view. Presumably we like to think that we enjoy *privileged access* to our own mental states. But if others could come to know telepathically what we're thinking or feeling, then (among other disquieting prospects) that would mean that our sins of the heart and most embarrassing or repulsive fleeting thoughts would potentially be available for public inspection.

But how well-founded is that belief or concern? To get a grip on the issues, we should begin by considering the valuable distinction (perhaps first mentioned by C. D. Broad [Broad, 1953, 1962] between telepathic (or clairvoyant) *cognition* and telepathic (or clairvoyant) *interaction*. As you would expect, every instance of the former would be an instance of the latter, but the converse doesn't hold—that is, ESP interaction may occur without ESP cognition. To see why this matters, we must take a closer look.

If telepathic cognition occurs at all, it would presumably be a form of non-sensorial *knowledge* about another individual's state of mind. More specifically, it would be a state of affairs in which so-called "percipient" *A* comes to know something about a telepathic interaction *A* has with another individual *B*. And what kind of things might *A* telepathically come to know? Well, presumably, in its most robust (and most intrusively intimidating) form, *A* would learn what's going on in *B*'s mind—that is, that *B* is having certain thoughts, perceptions, or emotions. But it would still be an instance of telepathic cognition—admittedly, less intimidating or threatening to one's mental privacy—if A learned merely that B was the telepathic cause of A's current thought or experience—that is, that B was directly influencing or interfering with A's stream of consciousness, whether or not A's resulting thoughts or experiences were those of B or known by A to be those of B.

However, Broad shrewdly recognized that the evidence for telepathy was seldom (if ever) evidence of these kinds of knowledge. On the contrary, what we usually find is evidence suggesting only telepathic *interaction*. In what is probably telepathy's most commonly reported form, person B's mental state *merely influences* that of A, and A learns nothing from the process about B's causal role, much less details of what B is thinking or feeling. For example, it would be telepathic interaction (not cognition) if my thought of the Eiffel Tower directly (that is, without sensory mediation) caused a remote person simply to think about the Eiffel Tower (or about towers generally, or about the Tower of London)—that is, without that person realizing that I played a causal role in that event, much less that I was thinking about the Eiffel Tower. Similarly, it would be a case of clairvoyant interaction (not cognition) if a burning house was the direct cause of someone at a remote location simply thinking about fire (or heat), or feeling a need to apply aloe to one's skin, or having the urge to watch *Blazing Inferno*. There's no need (and arguably not even a temptation) in these cases to insist that the percipient knows (presumably subconsciously) what caused the experience in question. The telepathic and clairvoyant scenarios would simply be paranormal analogues to the way our bombardment with environmental information can trigger various thoughts and associations, and perhaps distant or idiosyncratic associations at that. In both the paranormal and normal cases, we may be oblivious to the causal processes that led to our thoughts.

As it happens, when we look closely at the evidence for apparent telepathy, it does indeed seem as if it's largely (though perhaps not entirely) evidence merely of telepathic interaction. But we must make an important admission even before we look at the evidence—namely, that as far as we know, telepathy could occur between strangers or only very casual acquaintances, with the percipients never learning why, or even *that*, they had experienced telepathically influenced (or tainted) mental states. We have no grounds at present for denying this possibility, and, if that sort of telepathy occurs, we have no idea whether those moments of telepathic interaction are frequent or rare.

Granted, percipients in spontaneous cases (such as crisis cases) sometimes seem to know (or at least infer or suspect) who caused the surprising or anomalous thought they just had. And that's to be expected. After all, if I have an intrusive thought that my friend Jones had an accident and is in pain, it's a natural (though rationally risky) next move to infer that I'd been in touch psychically and momentarily with Jones. Nevertheless, it's often (if not usually) the case that percipients only learn some time after their experience, and *through normal channels of information*, that their anomalous mental states corresponded to the roughly contemporaneous thoughts or experiences of a remote individual in crisis. So in those cases at least, knowledge of that correspondence doesn't seem to be *telepathic* cognition.

In the interest of full disclosure, I must report that one could attempt here a theoretical maneuver that perhaps only a philosopher could love. The point of the maneuver is to argue that even in cases where there seems to be only mere telepathic interaction, what we find instead is a cornucopia of cognition. One could argue that the percipient's original telepathically caused mental state was indeed telepathic cognition—presumably subconscious. And then one could claim that the percipient's subsequent knowledge of the correspondence between the earlier telepathic experience and the agent's crisis is a form of *second-order knowledge*—that is, non-telepathic knowledge that the earlier mental state was an instance of telepathic knowledge. So one could claim that at the time of the original telepathic interaction, the percipient knows that Jones is (or was) in crisis but doesn't know that (s)he knows this.[1]

However, if the appeal to second-order knowledge is viable at all, it may be applicable only to crisis cases. More typically, correspondences between the thoughts of agent and percipient are less clear-cut, and don't seem at all to refer or point to a presumptive agent. So they don't require positing any telepathic awareness or cognition of the agent's causal role, much less what the agent's mental state is. For example, in one well-known experiment in ostensible dream telepathy, the agent was concentrating on a target-print of Bichitir's *Man with Arrows and Companion*, which portrays three men in India sitting outdoors. One holds a musical stringed instrument; the most prominent of the three holds a bow and arrows. The third man has a stick over his shoulder that looks like a rifle muzzle. One minor detail of the painting is a stake with a rope tied around it, and the percipient seemed to pick up on that small detail and incorporate images of rope prominently in his dreams. He had five dreams that night, and three of them contained rope (or coiled rope) as a prominent feature. Moreover, in another dream the percipient saw a "hammock in which there was an awful lot of suspended strings" (Ullman, Krippner, & Vaughan, 2002, p. 125).

In another study, the agent (an orthodox Jew) concentrated on a print of Chagall's *The Yellow Rabbi*, in which an old rabbi sits at a table with a book in front of him. The subject of the experiment was a Protestant. In one dream, he saw a man in his 60s riding in a car. In another he reported "a feeling of older people. The name of Saint Paul came into my mind." In another, he dreamt of a professor of humanities and philosophy reading a book. In the summary of his dreams the next morning, the subject reported, "So far, all I can say is that there is a feeling of older people. . . . The professor is an old man. He smoked a pipe, taught humanities as well as philosophy. He was Anglican minister or priest" (Ullman et al., 2002, pp. 91ff).

So when percipients are participating in informal experiments with friends or in more formal experiments (like the Ganzfeld), it may not be outrageous to say that they can know whose mental state affected their own. But if so, it's only because the percipients understand from the start, *and through normal channels of information*, that there's a designated agent (or "sender") and that the goal of the experiment is to find significant correspondences between the mental states of the agent and percipient. There doesn't seem to be even a superficial basis for saying that percipients had telepathic knowledge all along, but didn't know that they know.

In any case, there's another reason to question whether the percipients' conjectures in these situations are types of knowledge. Perhaps the following analogy will make this clear. Suppose an unidentified person surreptitiously deposits a message with my signature or photo on your doorstep. Obviously, the deposited object doesn't indicate unambiguously who put it there. After all, it could be left there mischievously by someone other than me. Knowledge of the object's source can't be derived simply from the object's presence on your doorstep. Similarly, a percipient's telepathically induced state won't point unambiguously to its source, even if it contains features that seem to "refer" or point to a source.

Besides, and as the dream-telepathy examples illustrate, a telepathically induced state needn't contain *any* such clues or pointers, and the vast majority of ostensible telepathic interactions lack those features. So there's no reason to think that paranormal experiences must include (or be preceded by) a warning or marker—something analogous to a flourish of trumpets, announcing that the experience is paranormal. Therefore, as long as percipients lack additional, normally acquired contextual information about the presumed origin of their ESP-induced mental state, that state might seem to be a merely random intrusive thought—that is, one of the occasional incongruous or unexpected, and easily ignored, mental states probably all of us have during the course of the day.

We should now see clearly one reason why the ESP cognition/interaction distinction matters. If (as it seems) most ostensible telepathy cases are examples of telepathic interaction but not telepathic cognition, then we may have no grounds for worrying about an ongoing (or at least significant) loss of mental privacy. And that may be enough for us to feel we're at least generally off the hook, and that we'll be able to shield our most reprehensible or embarrassing thoughts from prying minds.

Incidentally, this is one reason why the lamentably trendy practice of replacing the venerable terms "ESP," "telepathy," and "clairvoyance" with the catch-all term "anomalous cognition" (AC) is egregiously wrong-headed. (See, e.g., May, Spottiswoode, Utts, & James, 1995; May, Utts, & Spottiswoode, 1995a, 1995b.) I criticized that practice some time ago (Braude, 1998), and I've refined and expanded that critique in a forthcoming book (Braude, 2020). The same may be said about the even more recently trendy (and arguably incoherent) terms "nonlocal awareness" or "nonlocal consciousness." But what I hope the preceding has shown is that by ignoring the useful cognition/interaction distinction, those terminological reduction strategies fail to supply the taxonomic resources even for beginning to describe adequately the relevant, interesting, and empirically unresolved issues discussed above.

Let's pursue the matter a bit, because it should shed further light on the feasibility of positing telepathic cognition. Perhaps the clearest examples of mere telepathic interaction are those in which a person's mental states seem to be the direct cause of a remote individual's *actions*. As Jule Eisenbud noted,

> That a person can mentally influence not just the thoughts of other persons extrasensorially at a distance but also their decisions and actions must be one of the oldest facts of nature known to man. It has been woven into the core of every primitive culture described by anthropologists. (Eisenbud, 1992, p. 87)

But the evidence for this isn't simply anecdotal. In relatively modern times, the phenomenon has been investigated systematically and experimentally, and the best-documented cases concern the induction of hypnotic states at a distance. For example, hypnosis at a distance was reported in the eighteenth century by the early mesmerists, including Puységur, and then, in the mid to late nineteenth century, in studies by Janet and Richet (Janet, 1885, 1886; Richet, 1885, 1888; for more details, see Eisenbud, 1992). Perhaps defenders of the use of the term "anomalous cognition" forgot about this body of evidence, or more likely didn't know about it at all. Too often, psi researchers enter the field with—at best—only a very superficial knowledge of the rich history of relevant empirical and theoretical work that preceded them.[2] Nevertheless, it seems indefensible for partisans of the new terminology to exclude the phenomenon from their terminological considerations.

Interestingly, though, when it comes to the studies of apparent telepathic mind-control, even those familiar with the evidence do their best to avoid the subject. For example—and despite their successes—Janet and Richet abandoned the study of hypnosis shortly after completing their experiments and retreated to less momentously intriguing lines of investigation. Moreover, when Vasiliev demonstrated hypnosis at a distance once again in the mid twentieth century, the community of psi researchers (and of course the rest of the academy) failed to pursue the matter further.[3] In fact, and contrary to what usually happens when parapsychologists report much less dramatic and noteworthy effects, there was no flurry of replication attempts—actually, no attempts at all. It's not that Vasiliev's work (or that of his precursors) was poorly done. Rather, it seems clear that the phenomenon was simply terrifying in its implications and thus too easily ignored.

Partisans of the term "anomalous cognition" might be tempted to argue that the expression "cognition" was merely a terminological infelicity, suggesting (admittedly misleadingly) that every instance of AC is a kind of knowing or cognition. What matters, they might say, is that AC is merely a kind of anomalous "information transfer" or "acquisition of information." So they might suggest that some sort of information is acquired or transferred even when a thought about the Eiffel Tower causes someone to think about the Tower of London, or when a burning house causes someone to think about matches, or when someone remotely responds to my hypnotic command to fall asleep. But even if that terminological maneuver works for some instances of ESP interaction, other ostensibly telepathic and clairvoyant interactions more clearly resist definition or analysis in terms of information transfer.

At stake here is another intimidating issue, a modest extension of what we considered in connection with hypnosis at a distance— namely, that telepathic influence could—at least theoretically— be used for *total* control of another person's mind and body. Now presumably we wouldn't want to say that telepathic dominion over my thoughts and actions can be understood in terms of transfer or acquisition of information. After all, we wouldn't invoke information transfer to explain extreme, but normal, forms of forcing another to act. Whatever exactly the process might be, it's not analogous (say) to understanding and responding to a command. The clearest examples are probably ordinary cases of behavioral coercion. It's not information transfer, in any helpful *epistemic* sense of the term "information," if I physically overpower you and compel you to pull the trigger of a gun, and we similarly wouldn't consider it to be information transfer if my willing alone both prevented you from exercising your volition and also compelled you to fire the gun. Perhaps we should describe that telepathic version of coercion as a form of *possession*. But what matters is that the degree of control posited in these coercion scenarios resembles the control of a puppet, and it's thoroughly unilluminating to describe the puppeteer as transferring information to the puppet.

Likewise, we wouldn't consider it to be information transfer if I telepathically seized control of your mental life, blocking your access to your own stream of consciousness, and forcing you to have thoughts that are not your own. Victims of such telepathic influence would have no awareness at all—much less knowledge—of the interaction. And that's one reason why we wouldn't hold them morally responsible for their thoughts and actions at the time.

Before you dismiss these proposed scenarios as mere fantasy, we should note that there's actually an empirical basis for concern about this issue. It's not simply an abstract, theoretical matter we can acknowledge and then conveniently put out of mind. Total telepathic control of a human organism is ostensibly what happens during *mediumistic trance-impersonation*, in which the medium's body (and presumably, brain) are controlled by a deceased communicator who also apparently displaces the medium's waking consciousness. This is the process F. W. H. Myers called "telergy," and it remains an open question whether discarnate telepathic control is what really happens during mediumistic trance impersonations, or whether (say) it's the medium's dissociative dramatic personation instead, with occasional verifiable mediumistic ESP thrown in for good measure.[4] At any rate, if there's a bright side to the possibility of telergy, it's that the process doesn't seem to require or involve some dreaded form of "mind-reading" on the part of either agent or percipient. Rather, it would be a situation in which one individual's mental states *displace* another individual's ordinary stream of consciousness.

Partisans of "anomalous cognition" might be tempted to reply that telergy should properly be called "anomalous perturbation" (AP), or (in more virtuous language) "PK". But that would blur the admittedly somewhat fuzzy, but at least apparently useful, distinction between telepathic influence and PK. For all we know at our still preliminary level of understanding, the paradigmatic PK events of levitating a table, materializing a human figure, or biasing a random event generator, may be significantly different processes, *not only from each other* but also from directly influencing a person's thoughts or actions. So until we have good reason for claiming that all these phenomena can be similarly explained, it seems unwise at the very least to embrace terminology that prevents us from tentatively classifying the latter only as a distinct, telepathic process.

Actually, conflating telepathic influence and PK will probably appeal only to physicalists who would interpret the latter as a purely physical process and the former as a kind of physical influence on the percipient's brain. But if (as it seems) reductive physicalism is generally untenable, it again seems wise (for now at least) to entertain the possibility that telepathic influence and PK are distinct phenomena.

We're fortunate to have developed the linguistic resources for making fine distinctions between classes of phenomena whose differences certainly make sense in theory, and which also seem to have empirical warrant. If later empirical or theoretical advances show that our distinctions have no basis in fact and only apparent theoretical utility, we can then comfortably simplify our arsenal of parapsychological categories. But we're a long way from that point. In the meantime, then, the proposed taxonomical reform of replacing "ESP," "telepathy," and "clairvoyance" with "anomalous cognition" is unacceptably coarse, quite apart from the other serious shortcomings I enumerate elsewhere (Braude, 1998, 2020).

Now we can return to the issue of mental privacy. We've seen that there are quite diverse, and even unsettling, forms of apparent telepathic interaction without cognition. However, it remains an open question whether we have decent evidence of any form of telepathic cognition—especially of a kind that would justify fearing the loss of privileged access to our own mental states. If this Editorial has helped at all to clarify the issues, then perhaps we now have a better idea what sort of ostensibly telepathic evidence to look for. In the meantime, we can probably and comfortably continue living our unsavory private inner lives.

NOTES

1 We also can't rule out that the percipient's telepathically influenced experience occurs simultaneously with clairvoyant awareness of the crisis occurring to the agent. In that case, we should be reluctant to consider the incident a case purely of telepathic cognition—perhaps GESP [general ESP] cognition instead.

2 This is not entirely their fault. Parapsychology, unlike mainstream disciplines, offers few opportunities to undertake systematic and comprehensive study of psi before embarking on one's own research.

3 Vasiliev (1976). For a good discussion of telepathy at a distance, see

Eisenbud (1970, Chapter 5; 1992, Chapter 6).
4 See Braude (2003) for a discussion of these ideas.

REFERENCES

Braude, S. E. (1998). Terminological reform in parapsychology: A giant step backwards. *Journal of Scientific Exploration*, 12(1), 141–150.

Braude, S. E. (2003). *Immortal remains: The evidence for life after death*. Rowman & Littlefield.

Braude, S. E. (2020). *Dangerous pursuits: Mediumship, mind & music*. Anomalist Books.

Broad, C. D. (1953). *Religion, philosophy, and psychical research*. Routledge & Kegan Paul. [Originally published in *Philosophy*, 24 (1949), 291–309. doi:10.1017/S0031819100007452]

Broad, C. D. (1962). *Lectures on psychical research*. Routledge & Kegan Paul. [Reprinted by Routledge, 2011.]

Eisenbud, J. (1970). *Psi and psychoanalysis: Studies in the psychoanalysis of psi-conditioned behavior*. Grune & Stratton.

Eisenbud, J. (1992). *Parapsychology and the unconscious*. North Atlantic Books.

Janet, P. (1885). Note sur quelques phenomènes de somnambulisme. *Revue Philosophique de la France et de l'Etrangere*, 21, 190–198. [1968 translation, Report on some phenomena of somnambulism, *Journal of the History of the Behavioral Sciences*, 4(2), 124–131.]

Janet, P. (1886). Deuxième note sur le sommeil provoqué à distance et la suggestion mentale pendant l'état somnambulique. *Revue Philosophique de la France et de l'Etrangere*, 22, 212–223. [1968 translation, Second observation on sleep provoked from a distance and mental suggestion during the somnambulistic state. *Journal of the History of the Behavioral Sciences*, 4(3), 258–267.]

May, E. C., Spottiswoode, S. J. P., Utts, J. M., & James, C. L. (1995). Applications of decision augmentation theory. *Journal of Parapsychology*, 59(3), 221–250.

May, E. C., Utts, J. M., & Spottiswoode, S. J. P. (1995a). Decision augmentation theory: Applications to the random number generator database." *Journal of Scientific Exploration*, 9(4), 453–488.

May, E. C., Utts, J. M., & Spottiswoode, S. J. P. (1995b). Decision Augmentation Theory: Toward a model of anomalous mental phenomena. *Journal of Parapsychology*, 59, 195–220. https://www.scientificexploration.org/docs/9/jse_09_4_may.pdf

Richet, C. (1885). Un fait de somnabulisme à distance. *Revue Philosophique de la France et de l'Etrangere*, 21, 199–200.

Richet, C. (1888). Expériences sur le sommeil à distance. *Revue de l'hypnotisme*, 2, 225–240.

Ullman, M., Krippner, S., & Vaughan, A. (2002). *Dream telepathy: Experiments in nocturnal extrasensory perception*. Hampton Roads.

Vasiliev, L. L. (1976). *Experiments in distant influence: Discoveries by Russia's foremost parapsychologist*. Dutton. [Reprinted in 2002 as *Experiments in mental suggestion*, Hampton Roads.]

Journal of Scientific Exploration, Vol. 34, No. 2, pp. 209–232, 2020   0892-3310/20

RESEARCH ARTICLE

A New Model to Explain the Alignment of Certain Ancient Sites

Mark J. Carlotto

Submitted July 19, 2019; Accepted September 4, 2019; Published June 30, 2020

https://doi.org/10.31275/2020/1619
Creative Commons License CC-BY-NC

Abstract—In a previous study of more than two hundred ancient sites, the alignments of almost half of the sites could not be explained. These sites are distributed throughout the world and include the majority of Mesoamerican pyramids and temples that are misaligned with respect to true north, megalithic structures at several sites in Peru's Sacred Valley, some pyramids in Lower Egypt, and numerous temples in Upper Egypt. A new model is proposed to account for the alignment of certain unexplained sites based on an application of Charles Hapgood's hypothesis that global patterns of climate change over the past 100,000 years could be the result of displacements of the Earth's crust and corresponding shifts of the geographic poles. It is shown that more than 80% of the unexplained sites reference four locations within 30° of the North Pole that are correlated with Hapgood's hypothesized pole locations. The alignments of these sites are consistent with the hypothesis that if they were built in alignment with one of these former poles they would be misaligned to north as they are now as the result of subsequent geographic pole shifts.

Keywords: ancient sites; pyramid alignment; pole shifts

INTRODUCTION

In a previous study of ancient sites, the alignments of almost half of the sites could not be explained (Carlotto, 2020). These sites, which are distributed throughout the world, include the majority of Mesoamerican pyramids and temples that are misaligned with respect to true north, megalithic structures at several sites in Peru's Sacred Valley, some pyramids in Lower Egypt, and numerous temples in Upper Egypt. From a review of the archaeological and archaeoastronomical literature, eight basic reasons were identified that typically account for the orientation of an archaeological site: 1) to cardinal directions (i.e. facing north, south, east, and west), 2) to solstice sunrise or sunset directions, 3) to sunrise or sunset directions on days when the sun passes directly overhead, 4) to directions of major and minor lunar standstills, 5) to a planet, 6) to a star or constellation, 7) to magnetic north, and 8) in the direction of a site of religious or spiritual importance. We also considered other explanations such as landscape and topography that have been used in some cases to account for the alignment of certain sites. For example, Shaltout and Belmonte (2005) analyzed the orientation of more than one hundred temples in Upper Egypt and Lower Nubia to discover that they face many different directions. Their principal conclusion is that local topography (the course of the Nile), not astronomy, was the most important factor in aligning the foundations of the temples.

This paper proposes a new model to explain the alignment of certain sites throughout the world based on an application of Charles Hapgood's hypothesis that patterns of climate change over the past 100,000 years could be the result of displacements of the Earth's crust and corresponding shifts of the geographic poles. The next section discusses the origin of the idea of geographic pole shifts, how information about the motion of the geo-magnetic poles over time suggests that large shifts of geographic poles have occurred in the past, and possible relationships between geographic pole shifts and climate change. The following section describes a new model to explain the alignment of sites to previous pole locations based on an application and refinement of Hapgood's original pole shift hypothesis. Results are organized into eight geographic regions. It is shown that more than 80% of the unexplained sites in our previous study (Carlotto, 2020) reference at least one of these previous pole locations.

GEOGRAPHIC POLE SHIFTS AND CLIMATE CHANGE

Early in the 20th century, Alfred Wegener and others theorized that the continents were once a single large landmass that broke up and slowly drifted apart. Wegener's theory of continental drift explained the complementary shape of coastlines—how the west coast of Africa seems to fit the east coast of the Americas—and the similarity in rock formations and fossils along matching coastlines. This theory, now known as plate tectonics, divides the outer layer of the crust, called the lithosphere, into a number of plates that move independently of one another over a less rigid layer called the asthenosphere (Kious & Tilling, 1996). Holmes (1944) proposed that the Earth's mantle contains convection cells that dissipate interior heat and move the crust at the surface, thus providing a physical mechanism to drive plate motion. Inspired by Wegener's work, Milanković (1932) investigated the movement of the poles that he believed worked together with plate motion so that "the displacement of the pole takes place in such a way that . . . Earth's axis maintains its orientation in space, but the Earth's crust is displaced on its substratum."

The earth's axis of rotation intersects the surface at the north and south geographic poles, which are currently located in the Arctic and Antarctic. The flow of liquid metal in the outer core generates electric currents. The rotation of earth on its axis causes these electric currents to induce a magnetic field. The location of the magnetic poles slowly wanders in a seemingly random manner around the geographic poles. Rocks, sediment, and archaeological artifacts that contain magnetic minerals such as magnetite record the direction and intensity of Earth's magnetic field when they are heated above the Curie temperature. When a paleomagnetic material cools, magnetic information is retained by the mineral grains. By collecting and analyzing samples at different times and in different places, it is possible to estimate the location of the magnetic poles (paleopoles) as a function of time.

Kirschvink et al. (1997) determined from paleomagnetic data collected in Australia and North America that a massive crustal shift occurred between 534 million and 505 million years ago, which caused Australia to rotate a quarter of the way around the globe. This shift occurred around the time of the Cambrian Explosion, when most groups of animals first appear in the fossil record, and is thought to have been a major factor in the evolutionary changes that later took place. Woodworth and Gordon (2018) used paleomagnetic and ocean sediment data to show that Greenland was much closer to the North Pole 12–48 million years ago than it is today. Daradich et al. (2017) estimate a steady shift of Earth's poles by ~8° over the last 40 million years toward Greenland has brought North America to increasingly higher latitudes and caused its climate to gradually cool over this period of time.

If polar motion affects climate, the converse may also be true. Prior to the year 2000, the North Pole was slowly moving toward Hudson Bay, at which time it changed direction and began to drift toward Greenland. Chen at al. (2013) claim that the change in direction was caused by the accelerated melting of the Greenland Ice Sheet. Adhikari and Ivins (2016) argue that polar motion is influenced by changes in the amount of water held within the continents. Although these factors appear to control the direction of polar motion, they do not appear sufficient to account for its magnitude. Adhikari et al. (2018) have come to the conclusion that mantle convection, which drives plate tectonics, also seems to be a significant factor affecting polar motion.

It is generally assumed that climate patterns are driven to a large extent by the amount of solar radiation reaching the Earth. The amount of radiation depends on a combination of factors including changes in the eccentricity in our orbit around the sun, axial tilt (obliquity), axial and apsidal precession, and orbital inclination. The combination of these effects gives rise to what are called Milanković cycles. Although there is extensive evidence that the variation in solar radiation is an important factor, there are certain problems with Milanković's model as it relates to the timing and magnitude of the cycles and their correlation with climate events. Muller and MacDonald (1997) suggest the possibility that an external factor such as extraterrestrial accretion of dust or meteoroids could affect climate. It has been hypothesized that the Younger Dryas period of rapid cooling in the late Pleistocene, 12,800 to 11,500 years ago, could have had an extraterrestrial cause such as the Taurid meteor swarm (Napier, 2010). Woelfli et al. (2002) propose that an encounter with a Mars-sized object at around this time moved the North Pole from Greenland to its present position.

SHIFTED GEOGRAPHIC POLE SITE ALIGNMENT MODEL

Hapgood (1958) hypothesized that climate changes and ice ages could be explained by large sudden shifts of the geographic pole. He cites extensive evidence suggesting that during the last ice age the North Pole was located at around 60° N, 83° W, near Hudson Bay in Canada. Using climate data from a variety of sources, Hapgood reasoned that North America, which was then covered by a massive layer of ice and

TABLE 1
Estimated Locations of the North Pole

Name	Latitude	Longitude
Hudson Bay	59.75°	−78.0°
Greenland	79.5°	−63.75°
Norwegian Sea	70.0°	−0°
Bering Sea	56.25°	−176.75°

snow, was colder because it had been shifted closer to the pole, while places on the opposite side of the earth, such as Europe, were warmer because they had been shifted away from the pole and south toward the equator. By examining patterns of climate change, he estimated that three geographic pole shifts had taken place during the past 100,000 years: 1) from Hudson Bay (60°N 73°W) to the current pole, 12,000 to 17,000 years ago, 2) from the Atlantic Ocean between Iceland and Norway (72°N 10°E) to Hudson Bay, 50,000 to 55,000 years ago, and 3) from the Yukon (63°N 135°W) to between Iceland and Norway, 75,000 to 80,000 years ago.

Rand Flem-Ath noted that if the North Pole were in Hudson Bay, the major axis of Teotihuacan, an ancient Mesoamerican city 25 miles northeast of modern-day Mexico City, which is currently oriented 15.4° east of north, would be aligned to within a few degrees of due north (Wilson & Flem-Ath, 2000). Motivated by this observation, more than fifty sites not aligned to north were identified that could have once been aligned to one of Hapgood's hypothesized pole locations. An algorithm was developed that used the orientation (azimuth) angle and geographic coordinates of these sites measured in Google Earth to estimate a set of hypothetical "best-fit" pole locations (Carlotto, 2019). Table 1 lists the four hypothetical locations of the North Pole computed by this algorithm. The estimated Hudson Bay pole location is less than 200 miles from Hapgood's original Hudson Bay pole. If the North Pole were at that location, Teotihuacan would be aligned to the cardinal directions. The estimated pole in northern Greenland is 1,250 miles west of Hapgood's original Iceland/Norway pole, and the estimated pole in

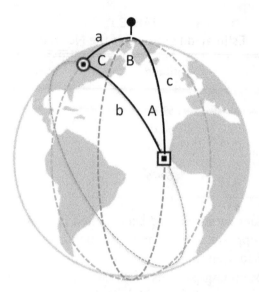

Figure 1. The locations of a site A, North Pole B, and previous pole C are the vertices of a spherical triangle. Segments of spherical triangles are great circles. The angle A is the azimuth of the previous pole location measured at the site.

the Norwegian Sea is about 300 miles south of it. A fourth computed pole location is in the Bering Sea north of the Aleutian Islands, about 1,500 miles from Hapgood's original Yukon pole.

With reference to Figure 1, let A be the location of a site, B the current location of the North Pole, and C the location of the North Pole at the time the site was first established. The angle A is the current alignment of the site with respect to north. A shift in the geographic pole causes both the latitude as well as the orientation of a site to change. If (λ_A, ϕ_A) and (λ_C, ϕ_C) are the latitudes and longitudes of the site and past pole in the current geographic reference frame, the orientation (rotation) angle of the site is

where

$$A = \sin^{-1}\left[\frac{\sin a \sin B}{\sqrt{1 - (\cos a \cos c + \sin a \sin c \cos B)^2}}\right] \quad (1)$$

$$\alpha = \frac{\pi}{2} - \lambda_c$$

$$c = \frac{\pi}{2} - \lambda_A \tag{2}$$

$$B = \varphi_C - \varphi_A$$

Its latitude prior to the pole shift would have been $90° - \lambda_C$ where

$$\lambda_C \ \cos^{-1}(\cos a \cos c + \sin a \sin c \cos B) \tag{3}$$

By comparing the orientation angle of a site measured using Google Earth to Equation 1, it is possible to determine if the site could have once faced north. In addition, by substituting previous pole values from Equation 1 and Equation 3 into the solar and lunar alignment equations (Carlotto, 2020), it is possible to determine if the site was aligned to solstices, zenith passages, or lunar standstills relative to those poles.

RESULTS: SITES ALIGNED TO PREVIOUS POLE LOCATIONS

Tables 2–9 indicate the alignments for more than two hundred ancient sites to the current (Arctic Ocean) pole, and former estimated Hudson Bay, Greenland, the Norwegian Sea, and Bering Sea pole locations. The sites are organized into eight geographic regions. The key to the alignments is as follows:

> Cardinal directions, i.e., geographic poles, and equinoxes (E)
> Magnetic pole at the time of construction (X)
> Zenith passage (Z)
> Solstices (S)
> Major and minor lunar standstills (M,m)
> Stellar alignments (st)
> Alignments to "Sacred Directions" (D)

Only six of the eight alignment hypotheses were examined for the shifted poles, as there is insufficient information to evaluate "st", and "D" would not be affected by a crustal displacement.

TABLE 2
Alignments of Sites in Africa

Name	Latitude	Longitude	North	East	Current	Hudson Bay	Greenland	Norway Sea	Bering Sea
Algeria, Jabal Lakhdar	35.063404	1.183731	-5	85				E	
Egypt, Abu Rawash, Pyramid of Djedefre	30.032262	31.074714	0	90	E				
Egypt, Abusir, Pyramid of Neferefre	29.89377	31.201454	0	90	E				
Egypt, Abusir, Pyramid of Neferirkare	29.895093	31.202249	0	90	E				
Egypt, Abusir, Pyramid of Sahure	29.897622	31.203367	0	90	E				
Egypt, Abydon, Temple Ramses II	26.186426	31.91628	44.2	134.2				S	
Egypt, Abydos, Osirion	26.184099	31.918465	36.3	126.3		S			
Egypt, Abydos, Pyramid of Ahmose I	26.175056	31.937822	36	126		S			
Egypt, Abydos, Temple Seti I	26.184968	31.919183	36.3	126.3		S			
Egypt, Cairo, Mosque of Ibn Tulun	30.028691	31.249394	-39	51					
Egypt, Dahshur Pyramid of Senusret III	29.818888	31.22555	0	90	E				
Egypt, Dahshur, Bent Pyramid	29.790449	31.209324	0	90	E				
Egypt, Dahshur, Pyramid of Amenemhat II	29.805807	31.223038	0	90	E				
Egypt, Dahshur, Red Pyramid	29.808882	31.206113	0	90	E				
Egypt, Deir Bahari, Mortuary Temple of Mentuhotep II	25.737375	32.606178	23.2	113.2	S				
Egypt, Deir el Medinah, Temple of Hathor	25.728846	32.602128	-40	50			S		
Egypt, Dendera, Sacred Lake	26.14180698	32.66953166	16.1	106.1					E
Egypt, Dendera, Temple of Hathor	26.141914	32.670205	18.9	108.9	st,m			S	
Egypt, Edfu Temple of Horus	24.976747	32.873087	12.8	102.8				M	
Egypt, Elephantine, Temple of Khnum	24.084492	32.886206	-42	48			M		
Egypt, Giza, Khafre	29.975726	31.1308	0	90	E				
Egypt, Giza, Khufu	29.979067	31.13404	0	90	E				
Egypt, Giza, Menkaure	29.975811	31.131242	0	90	E				
Egypt, Kom Ombo	24.452085	32.928353	43.3	133.3		m		S	
Egypt, Lisht, Pyramid of Amenemhat I	29.574802	31.225304	0	90	E				
Egypt, Lisht, Pyramid of Senusret I	29.56016	31.22113	0	90	E				
Egypt, Luxor West, Temple Ramses III	25.719683	32.600711	-42	48			M		
Egypt, Luxor, Karnak, Temple of Amun Re	25.718484	32.659044	26.6	116.6	S				
Egypt, Meidum Pyramid	29.388368	31.157503	0	90	E				
Egypt, Pyramid of Teti	29.875142	31.221847	-12.5	77.5			E		
Egypt, Saqqara, Mastaba of Shepseskaf	29.838852	31.215273	0	90	E				
Egypt, Saqqara, Pyramid of Djedkare-Isesi	29.850983	31.220924	0	90	E				
Egypt, Saqqara, Pyramid of Djoser	29.87139735	31.21653162	5	95					
Egypt, Saqqara, Pyramid of Khendjer	29.832363	31.224043	0	90	E				
Egypt, Saqqara, Pyramid of Pepi II	29.840246	31.213496	0	90	E				
Egypt, Saqqara, Pyramid of Qakare Ibi	29.84159	31.217712	-10	80			E		
Egypt, Saqqara, Pyramid of Unas	29.868182	31.215012	0	90	E				
Egypt, Saqqara, Pyramid Userkaf	29.873332	31.219334	0	90	E				
Egypt, Shunet El Zebib	26.18951	31.908055	-41.7	48.3				m	
Egypt, Siwa Oasis, Amun Temple	29.201375	25.516151							E
Egypt, Temple of Edfu	24.978092	32.873475	3	93					
Egypt, Temple of Esna	25.29344448	32.55612504	-23	67	M				
Egypt, Temple of Hathor, El Kab	25.138586	32.828651	-44	46			M		
Egypt, Temple of Isis at Shenhur	25.86104	32.776808	10	100				M	
Egypt, Temple of Ramses II	25.727588	32.610283	41	131				S	
Egypt, Zawyet El Aryan, Layer Pyramid	29.93282	31.161262	-12	78			E		
Ethiopia, Bete Giyorgis	12.031714	39.04119	5.8	95.8	m				
Ethiopia, Yeha Temple	14.28570335	39.01911389	11.4	101.4				m	
Sudan, Dangeil, Amun Temple	18.131307	33.9598	16.5	106.5			S		E

E = cardinal directions, i.e. geographic poles and equinoxes. M, m = major and minor lunar standstills. S = solstices. st = stellar alignments. If no alignment is given, the reason is unknown. In some cases, there may be more than one explanation for an alignment.

TABLE 3
Alignments of Sites in Asia

Name	Latitude	Longitude	North	East	Current	Hudson Bay	Greenland	Norway Sea	Bering Sea
Cambodia, Koh Ker	13.78322	104.5374528	-12.5	77.5	Z				
Cambodia,Preah Khan of Kompong Svay	13.40382	104.75421	-28.2	61.8	M				
China, Chongling Mausoleum of Emperor Dezong of Tang	34.70738	108.82853	-4.2	85.8	x				
China, Jinling Mausoleum of Emperor Xianzong of Tang	34.570992	108.265923	-9	81	x				
China, The Lianhu Altar	36.632869	101.746123	15.8	105.8	S				
China, Tomb of Consort Ban	34.379801	108.704492	-11	79					
China, Tomb of Emperor Ai of Han	34.400855	108.764606	0	90	E				
China, Tomb of Emperor Cheng of Han	34.374896	108.698001	-10	80	x				
China, Tomb of Emperor Gaozu of Han	34.434691	108.876647	-14	76	x				
China, Tomb of Emperor Hui of Han	34.422895	108.841317	-17	73	x				
China, Tomb of Emperor Jing of Han	34.443823	108.940784	0	90	E				
China, Tomb of Emperor Ping of Han	34.397774	108.712421	0	90	E				
China, Tomb of Emperor Wen of Sui	34.28785	108.02289	-3	87	x				
China, Tomb of Emperor Wu of Han	34.338085	108.569684	-8	82	x				
China, Tomb of Emperor Xuan of Han	34.181063	109.022312	0	90	E				
China, Tomb of Emperor Yuan of Han	34.390303	108.739114	0	90	E				
China, Tomb of Emperor Zhao of Han	34.361753	108.640108	-11	79	x				
China, Tomb of Empress Dou	34.235825	109.118614	22.6	112.6	m				
China, Tomb of Empress Dowager Bo	34.220993	109.096341	21.6	111.6	m				
China, Tomb of Empress Fu	34.402608	108.772545	-4	86	x				
China, Tomb of Empress Li	34.340327	108.562002	-9.5	80.5	x				
China, Tomb of Empress Lü	34.433824	108.881292	-10.2	79.8	S				
China, Tomb of Empress Shangguan	34.363135	108.630538	-8	82	x				
China, Tomb of Empress Wang (a)	34.393242	108.733835	0	90	E				
China, Tomb of Empress Wang (b)	34.446291	108.9475	0	90	E				
China, Tomb of Empress Wang (c)	34.178951	109.028396	0	90	E				
China, Tomb of Empress Xu (a)	34.374648	108.68474	-9.5	80.5	x				
China, Tomb of Empress Xu (b)	34.12734	109.055786	0	90	E				
China, Tomb of Empress Zhang Yan	34.423195	108.836961	-15	75					
China, Tomb of Marquis Zhang Ao	34.427745	108.851209	-15	75					
China, Tomb of Princess Chengyang of Emperor Taizong	34.6156	108.49314	-6	84	x				
China, Tomb of Princess Xincheng of Emperor Taizong	34.62365	108.49888	-21	69					
China, Yarnaz Valley,Yarkhoto	42.952022	89.061138	-40	50	M				
India, Amritsar, Golden Temple	31.619938	74.876511	33.2	123.2	M				E
India, Chidambaram, Chidambaram Nataraja	11.399234	79.693715	-1	89	E				
India, Chitoor, Srikalahasti Temple	13.749686	79.698308	0	90	E				
India, Kanchipura, Ekambareswarar Temple	12.847302	79.699525	18.3	108.3	m				
India, Khadirbet, Dholavira	23.88690735	70.21377639	-5	85					
India, Madhya Pradesh, Sas Bahu Temple	16.018856	75.881959	-4	86			Z		
India, Madhya Pradesh, Tigawa Temple	23.690196	80.066918	-10	80			E		
India, Mahabalipuram, Shore Temple	12.616492	80.199267	13	103	Z	S			
India, Rameshwar Mandir	16.21768	73.462012	-14	76		E			
India, Shri Martand Sun Temple	33.745588	75.220286	-13.9	76.1		E			
India, Sigiriya	7.957173	80.760031	8.3	98.3	Z				S
India, Tamil Nadu					E,S,Z				
India, Thanjavur, Brihadisvara Temple	10.782614	79.131735	-20.5	69.5	st,m				
India, Tiruvannamalai, Annamalaiyar Temple	12.231884	79.06679	11.4	101.4	Z				
India, Udaipur Rajasthan, Sas Bahu Temple	24.735191	73.716283	-16	74		E			
India, Venkateswara Temple	13.68325	79.347195	-7	83			E		
Indonesia, Gunung Padang	-6.994518	107.056383	-20	70				E	
Inner Mongolia, Xanadu	42.356388	116.184304	0	90	E				
Japan, Osaka Castle	34.687298	135.525826	5.7	95.7			E		
Maldives, Thinadhoo	0.530107	72.99717	43	133	D				
Pakistan, Harappa	30.628104	72.863909	0	90	E				
Russia Por-Bazhyn	50.615271	97.384872	9.5	99.5				S	
Thailand Angkor Wat	13.412469	103.866986	0	90	E				
Thailand, Ayutthaya, Wat Phra Mahathat	14.356943	100.567509	-5.3	84.7	x				
Thailand, Kao Klang Nai, Sri Thep	15.465521	101.144681	9.5	99.5			Z		
Thailand, Prasat Hin Phimai	15.22093	102.493861	-22	68				E	
Thailand, Prasat Mueang Tam	14.496089	102.982608	-11	79				Z	
Thailand, Prasat Phanom Rung	14.532044	102.940223	-5.5	84.5	x				
Thailand, Prasat Si Khoraphum	14.944574	103.798352	0	90	E				
Thailand, Wat Phra Sri Rattana Mahathat	14.798673	100.613862	0	90	E				

D = "sacred directions". E = cardinal directions, i.e. geographic poles and equinoxes. M,m = major and minor lunar standstills. S = solstices. st = stellar alignments. X = magnetic pole at the time of construction. Z = zenith passage. If no alignment is given, the reason is unknown. In some cases, there may be more than one explanation for an alignment.

TABLE 4
Alignments of Sites in Europe

Name	Latitude	Longitude	North	East	Current	Hudson Bay	Greenland	Norway Sea	Bering Sea
Bosnia, Pyramid of the Sun	43.977259	18.176514	8.4	98.4					E
Greece, Athens, The Parthenon	37.971517	23.72659	-13.5	76.5			E		
Greece, Delphi Amphitheater	38.482477	22.500577	-38.2	51.8	D	E			
Greece, Knossos	35.297863	25.163092	11.8	101.8			m		E
Greece, Mycenae, Lion Gate	37.73075184	22.7564996	-40	50		E			
Greece, Mycenae, Tomb of Agamemnon	37.726725	22.754367	10.5	100.5			m		E
Greece, The Temple of Artemis	37.949611	27.363921	21	111					
Italy, Rome, Circus Maximus	41.885944	12.485215	36.7	126.7	M				
Italy, Rome, Palantine Hill	41.889209	12.487459	36.7	126.7	M				
Italy, Sardinia, Monte d'Accoddi	40.79075445	8.448907568	9.1	99.1				M	Z
Malta, Gozo, Ġgantija Temple	36.04726	14.269015	37	127	M				
Spain, Mosque-Cathedral of Cordoba	37.878906	-4.779387	-30.4	59.6	S				
Spain, Naveta d'Es Tudons	40.00307541	3.891652768	-19.2	70.8	m				
Turkey, Hagia Sophia	41.01314018	28.98318202	34.3	124.3	S				
Turkey, Hattusa	40.01994347	34.61545489	38	128	M	S		m	m
UK, Calanais Standing Stones	58.197566	-6.745127							
UK, Glastonbury Tor	51.144444	-2.698611	-26.5	63.5	m				
UK, Stonehenge	51.178868	-1.826163			S,M,m				

D = "sacred directions". E = cardinal directions, i.e. geographic poles and equinoxes. M,m = major and minor lunar standstills. S = solstices. If no alignment is given, the reason is unknown. In some cases, there may be more than one explanation for an alignment.

TABLE 5
Alignments of Sites in North America

Name	Latitude	Longitude	North	East	Current	Hudson Bay	Greenland	Norway Sea	Bering Sea
Canada, AB, Badlands Guardian	50.01037	-110.113133			E	.			
US, California, Blythe Intaglios, B1	33.800585	-114.532055	0	90	E				
US, California,Blythe Intaglios, B3	33.800402	-114.538078	29	119		E			
US, Georgia, Ocmulgee National Monument	32.838868	-83.606114	34	124	M				
US, New Mexico, Chaco Canyon, Pueblo del Arroyo	36.060854	-107.9663	24	114	M				
US, Ohio, Great Serpent Mound	39.02642	-83.431091	27.7	117.7	S				
US. Illinois, Cahokia, Monks Mound	38.660158	-90.062466			S,M				

E = cardinal directions, i.e. geographic poles and equinoxes. M = major lunar standstills. S = solstices. In some cases, there may be more than one explanation for an alignment.

TABLE 6
Alignments of Sites in the Pacific Ocean

Name	Latitude	Longitude	North	East	Current	Hudson Bay	Greenland	Norway Sea	Bering Sea
Chile,Easter Island, Ahu Akivi	-27.115014	-109.395043	-2.7	87.3	E				
Chile,Easter Island, Ahu Nau Nau	-27.074425	-109.322455	-19.6	70.4	m				
Chile,Easter Island, Ahu Tahai	-27.140076	-109.427314	8.3	98.3			E		
Chile,Easter Island, Ahu Tongariki	-27.125774	-109.276933	30	120	S				
Chile,Easter Island, Ahu Vinapu	-27.174098	-109.405737	8.1	98.1			E		
Micronesia, Nan Madol	6.844537	158.335795	-33	57	M		E		
Micronesia, Nan Madol, Temple of Nan Dawas	6.844537	158.335795	7	97	Z		E		
Samoa, Pulemelei Mound	-13.735237	-172.324399	-7.3	82.7					
Tonga, Ha'amonga 'a Maui Trilithon	-21.136606	-175.048087	32.7	122.7		E			

E = cardinal directions, i.e. geographic poles and equinoxes. M,m = major and minor lunar standstills. S = solstices. Z = zenith passage. If no alignment is given, the reason is unknown. In some cases, there may be more than one explanation for an alignment.

TABLE 7
Alignments of Sites in the Middle East

Name	Latitude	Longitude	North	East	Current	Hudson Bay	Greenland	Norway Sea	Bering Sea
Iran, Chogha Zanbil	32.00899687	48.5215934	-43.5	46.5		m	M	m	S
Iraq, Dur-Kurigalzu	33.35367069	44.20216381	-39.6	50.4				S	M
Iraq, Tower of Babel	32.536284	44.420803	-11.3	78.7			E		
Iraq, Ziggurat of Ur	30.962711	46.103126	-33.3	56.7	M				
Jerusalem, Dome of the Rock	31.778087	35.235306	-7.3	82.7	D				
Jerusalem, Western Wall	31.776657	35.23447	-12.1	77.9			E		
Jordan, Petra, Temple of the Winged Lions	30.330297	35.442554	17.5	107.5					E
Jordan, Qasr Il-Abd, Irak Al-Amir	31.912785	35.751941	-15	75				E	
Jordan, Umayyad Mosque in Amman	33.51159288	36.3066567	-6.4	83.6					Z
Lebanon, Baalbek, Temple of Jupiter	34.006694	36.203826	-12.2	77.8			E		
Saudi Arabia, Mecca, Kaaba	21.42251	39.826174	-34.9	55.1	M				
Turkey, Harran	36.865021	39.031565	9.6	99.6				m	
Yemen, Great Mosque of Sana'a	15.353123	44.214876	-25	65	M				

D = "sacred directions." E = cardinal directions, i.e. geographic poles and equinoxes. M,m = major and minor lunar standstills. S = solstices. Z = zenith passage. In some cases, there may be more than one explanation for an alignment.

TABLE 8
Alignments of Sites in South America

Name	Latitude	Longitude	North	East	Current	Hudson Bay	Greenland	Norway Sea	Bering Sea
Bolivia, Chincana Labyrinth	-15.990127	-69.202952	44	134	D			S	
Bolivia, Puma Punku	-16.56172	-68.680046	2	92			E		
Bolivia, Quenuani	-16.259407	-69.17127	-20	70	S			S	
Bolivia, Tiwanaku	-16.554933	-68.673487	2	92			E		
Peru, Caral-Supe	-10.893458	-77.52054	19.5	109.5	S				
Peru, Caral-Supe, Huanca Pyramid	-10.893458	-77.52054	19.5	109.5				E	
Peru, Chan Chan	-8.103554	-79.07076	19.5	109.5	m			E	
Peru, Chavin	-9.594527	-77.177002	14.7	104.7	D				
Peru, Cuzco	-13.518587	-71.975952							E
Peru, Huanuco Pampa	-9.875388	-76.816395	0	90	E				
Peru, Huayna Picchu, Temple of the Moon	-13.151931	-72.546507			M				
Peru, La Centinela	-13.45007514	-76.17223285	5.6	95.6				Z	
Peru, Machu Picchu, Temple of the Three Windows	-13.163592	-72.545414	-34.7	55.3					E
Peru, Machu Picchu, Terraces	-13.164219	-72.544831	-25	65			S		
Peru, Marcahuasi, Face	-11.77567	-76.581853	43	133	D			S	
Peru, Nazca Lines	-14.712825	-75.17485	19.3	109.3	D			E	E
Peru, Ollantaytambo, Temple of the Sun	-13.257536	-72.267129	-35	55					E
Peru, Sacsahuaman	-13.50933	-71.980916			D				
Peru, Sechin Bajo	-9.4648088	-78.26525923	-25.5	64.5	S				
Peru, Warawtampu	-10.46549	-76.536647	-24.2	65.8	D				
Peru, Chotuna	-6.720363	-79.952796	0	90	E				

D = "sacred directions." E = cardinal directions, i.e. geographic poles and equinoxes. M,m = major and minor lunar standstills. S = solstices. Z = zenith passage. In some cases, there may be more than one explanation for an alignment.

Of the 95 unexplained sites identified in our initial study, the shifted pole model is able to explain all but 17 of the alignments. 62 sites face one of the previous pole locations, 21 to solstices, and 21 to lunar standstills that reference previous pole locations. In some cases a site had more than one alignment; e.g., Knossos appears to be both

TABLE 9
Alignments of Sites in Mesoamerica

Name	Latitude	Longitude	North	East	Current	Hudson Bay	Greenland	Norway Sea	Bering Sea
Belize, Altun Ha, Sun God Pyramid	17.76395	-88.347061	7.6	97.6		E			
Belize, Xunantunich	17.088922	-89.141631	-10.3	79.7	D				
El Salvador, Tazumal	13.979547	-89.674131	18	108	m				
Guatemala, Mixco Viejo	14.871668	-90.664167	12.5	102.5	D	E			
Guatemala, Tikal	17.222094	-89.623614	8.6	98.6		E			
Guatemala, Yaxchilan	16.899655	-90.967093	30.4	120.4	D				
Honduras, Copan, Step Pyramids	14.84	-89.14			Z				
Mexico, Acatitlan	19.55	-99.17	20.3	110.3	Z			E	
Mexico, Alta Vista	23.478544	-103.945607				S,M,m			
Mexico, Bonampak	16.704	-91.065	38	128			M		
Mexico, Calakmul	18.105392	-89.810829	8.8	98.8		E			
Mexico, Calixtlahuaca	19.335038	-99.69757	-30	60	M				
Mexico, Chalcatzingo	18.676715	-98.770783	6.8	96.8			E		
Mexico, Chichen Itza	20.68	-88.57	21	111	M,Z			E	
Mexico, Chimalacatlan, C1	18.446236	-99.105878	-34.7	55.3	M				E
Mexico, Chimalacatlan, C2	18.444804	-99.104331	28.7	118.7	m				S
Mexico, Cholula	19.0583048	-98.30190553	25	115	S				
Mexico, Coba, Grand Pyramid	20.492974	-87.724195	-39	51		M			
Mexico, Comalcalco	18.27819958	-93.20032665	24	114	S				
Mexico, Cuauhtinchan Archeological Site, Cuauhcalli	18.9535	-99.502888	15.4	105.4	m				
Mexico, Cuicuilco	19.301021	-99.183798							
Mexico, El Cerrito Archaological Zone	20.551376	-100.444027	7.4	97.4			E		
Mexico, El Tajin, Pyramid of the Niches	20.448058	-97.378242	14.5	104.5		E			
Mexico, El Tajin, Southern Ballcourt	20.448058	-97.378242	0	90	E				
Mexico, El Tajin, Tajin Chico	20.448058	-97.378242	40	130					
Mexico, El Tepozteco	19.00078611	-99.1015579	26	116	m				
Mexico, La Venta	18.103191	-94.040946	-12.2	77.8	D				
Mexico, Mayapan	20.629823	-89.46059							
Mexico, Mitla	16.92704923	-96.35934812	12	102		E			
Mexico, Monte Alban	17.042122	-96.768184	6.45	96.45			E		
Mexico, Monte Alban, Building J	17.042122	-96.768184	-43	47	st				
Mexico, Palenque, North Group	17.483978	-92.04632	10.1	100.1		E			
Mexico, Palenque, Temple of the Inscriptions	17.48	-92.05	20.6	110.6				E	
Mexico, Tenango	19.108425	-99.597693	14	104		E			
Mexico, Tenochtitlan	19.435	-99.131389	7	97			E		
Mexico, Teotihuacan	19.6925	-98.843889	15.6	105.6	st	E			
Mexico, Tlatelolco	19.450994	-99.13751	8.5	98.5					
Mexico, Tula	20.064451	-99.3405	15.47	105.47	m				
Mexico, Tulum	20.21	-87.43	22.3	112.3				E	
Mexico, Uxmal, Palace of the Governors	20.359444	-89.771389	30	120	S				
Mexico, Uxmal, Pyramid of the Magician	20.359444	-89.771389	9.2	99.2		E			
Mexico, Uxmal, Templo Mayor	20.359444	-89.771389	19.6	109.6	Z			E	
Mexico, Xochicalco, Grand Pyramid	18.803889	-99.295917	0	90	E				
Mexico, Xochicalco, Temple of Quetzalcoatl	18.803889	-99.295917	15.4	105.4		E			

D = "sacred directions." E = cardinal directions, i.e. geographic poles and equinoxes. M,m = major and minor lunar standstills. S = solstices. st = stellar alignments. Z = zenith passage. If no alignment is given, the reason is unknown. In some cases, there may be more than one explanation for an alignment.

cardinally aligned to the Bering Sea pole and aligned to minor lunar standstills relative to the Greenland pole.

Figure 2 shows 4 of the 18 sites found that face the Hudson Bay pole. All of Teotihuacan is aligned to the Hudson Bay pole, as are structures in Tikal's North Acropolis. The Sri Martand Sun Temple in India does not currently face the sun, but if it were originally built when the North Pole was in Hudson Bay it would have been aligned as many sun temples are to the cardinal directions at that time. An extended area on

A) Teotihuacan, Mexico

B) Tikal, Guatemala

C) Sri Martand Sun Temple, India

D) Ha'amonga 'a Maui

Figure 2. Examples of sites aligned to the Hudson Bay pole. Photo credit: Apple Maps.

the island of Tonga surrounding the Ha'amonga 'a Maui Trilithon also is oriented in the direction of the Hudson Bay pole.

Twenty sites were found that face the Greenland pole. Four of the sites are shown in Figure 3. The Temple of Jupiter at Baalbek was built by the Romans over an earlier pre-Roman structure (Lohmann, 2010). Similarly, the Parthenon atop the Acropolis in Athens was built over an earlier Parthenon (Beard, 2010). Hannah (2013) reviews Dinsmoor's analysis of the Parthenon and concludes that on August 31, 488 BCE, Athena's "birthday," the sun would have risen north of east along the main axis of the temple. But how do we know when Athena, a goddess,

A) Tower of Babel, Babylon

B) Temple of Jupiter, Baalbek, Lebanon

C) The Parthenon, Athens

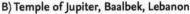

D) Tenochtitlan, Mexico City

Figure 3. Examples of sites aligned to the Greenland pole. Photo credit: Apple Maps.

the daughter of Zeus, was born? Was the Parthenon aligned with the sunrise on Athena's birthday, or was the date of Athena's birthday established based on the preexisting orientation of the Parthenon? And was this orientation, along with other structures on the Acropolis, originally directed toward an ancient pole in Greenland?

Chichen Itza and El Tepozteco in Mexico, Caral Supe in Peru, and the Brihadisvara Temple in India are 4 of the 12 sites found that face the Norwegian Sea pole (Figure 4). Southwest of the Temple of Quetzalcoatl at Chichen Itza, the Caracol is a dome-shaped structure thought to have been an observatory aligned to celestial events, including the

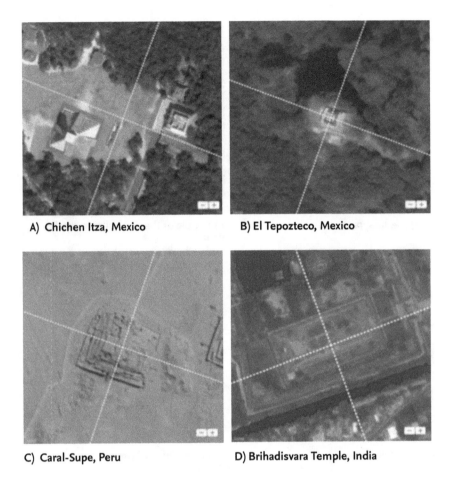

A) Chichen Itza, Mexico   B) El Tepozteco, Mexico

C) Caral-Supe, Peru   D) Brihadisvara Temple, India

Figure 4. Examples of sites aligned to the Norwegian Sea pole. Photo credit: Apple Maps.

summer and winter solstice sunrises and sunsets and the setting of the planet Venus. If this were its intended purpose, why are the Caracol, as well as the Temple of Quetzalcoatl and numerous other structures at Chichen Itza, all oriented in a decidedly non-solar direction, approximately 21.5° east of north, in the direction of the Norway pole?

Figure 5 shows 4 of the 12 sites that have been found to be aligned to the Bering Sea pole. One of the major Nazca lines is in the direction of the pole (another is in line with the Norwegian Sea pole). The Temple of the Sun at Ollantaytambo in Peru and the Temple of the Three Windows at Machu Picchu are aligned to the Bering Sea pole as are Knos-

A) Nazca line, Peru

B) Temple of the Sun, Ollantaytambo, Peru

C) Knossos, Crete

D) Temple of the Winged Lions, Petra, Jordan

Figure 5. Examples of sites aligned to the Bering Sea pole. Photo credit: Apple Maps.

sos in Crete and the Temple of the Winged Lions in Petra, Jordan. The direction of the Bering Sea pole is also closely aligned with a pattern of lines called ceques emanating out from the City of Cuzco.

In addition to sites aligned to the cardinal directions, the shifted pole alignment model accounts for 42 previously unexplained sites that could once have been aligned to solstices and to lunar standstills. Four of the sites are shown Figure 6. The Osireon in Abydos is thought to have been an integral part of Seti I's funerary temple yet it was originally built at a considerably lower level than the foundations of the temple (Petrie & Murray, 1903). It is not currently aligned in any direction of

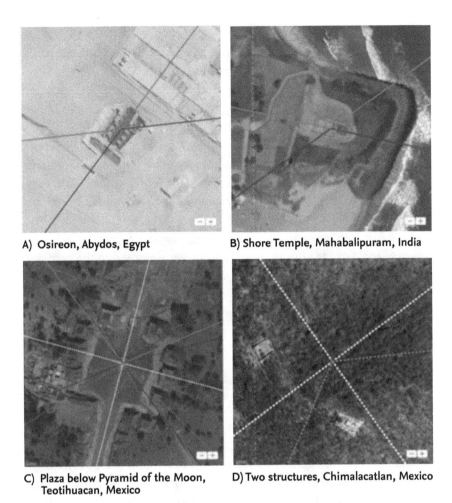

A) Osireon, Abydos, Egypt

B) Shore Temple, Mahabalipuram, India

C) Plaza below Pyramid of the Moon, Teotihuacan, Mexico

D) Two structures, Chimalacatlan, Mexico

Figure 6. Examples of other sites that reference previous pole locations. Pairs of solid lines are the summer and winter solstice alignments. Dotted lines are lunar standstill directions. **A)** is aligned to the summer solstice sunrise/winter solstice sunset relative to the Hudson Bay pole. **B)** is aligned to the winter solstice sunrise/summer solstice sunset relative to the Hudson Bay pole. **C)** shows solar and lunar alignments relative to the Hudson Bay pole. **D)** is aligned to cardinal directions and major lunar standstills relative to the Bering Sea pole. Photo credit: Apple Maps.

astronomical significance. According to our proposed model, the Osireon would be aligned to solstices if the North Pole were in Hudson Bay. This is also the case for the Shore Temple in Mahabalipuram, India. At Teotihuacan, the Pyramid of the Moon, the Pyramid of the Sun, the

Temple of Quetzalcoatl, and the Avenue of the Dead all are aligned in the direction of the Hudson Bay pole. Numerous other structures in the plaza south of the Pyramid of the Moon that do not now reference any obvious astronomical events would have been aligned to solstices and lunar standstills at that time. Structures at Chimalacatlan in Mexico (Vigato, 2015) also appear to reference the Bering Sea pole.

Figure 7 plots the distribution of site alignments within the eight geographic regions versus pole location. Although the sample size is somewhat limited, certain patterns are evident. There are far more sites in Africa and Asia that are currently aligned to the cardinal directions than to any other direction at any other time. Over time, the number of sites appears to have increased in Mesoamerica and decreased in South America. Most of the sites in these regions were aligned to the cardinal directions. On the other hand, most sites in Europe and the Middle East were aligned to the moon. Where sites exist from the time of the Bering Sea pole to the present in six of the eight regions, there are no sites in the Pacific before the Greenland pole or in North America before the Hudson Bay pole.

DISCUSSION

Aveni (1980) states that modern cities tend to be built over the sites of earlier settlements, often preserving the original alignments for convenience of construction, and notes that the alignments of churches and planted fields in certain regions of Mexico follow the directions of alignments that had already been established in pre-Columbian times. Our hypothesis is that, over time, as certain sites fell into ruin, they were rebuilt and expanded, and new structures were added above and/ or around them consistent with the original site plan. What remains today thus indicates the original alignment of the site. In sites that contain both cardinally aligned and non-aligned structures there might not be obvious differences between the two if the older non-aligned structures were rebuilt or built over at the same time new cardinally aligned structures were added. Perhaps deeper excavations at these sites would provide further evidence of the antiquity of the non-aligned structures.

The proposed shifted geographic pole alignment model explains many archeoastronomical enigmas including the distributions of temple and pyramid alignments in Mesoamerica and previously unex-

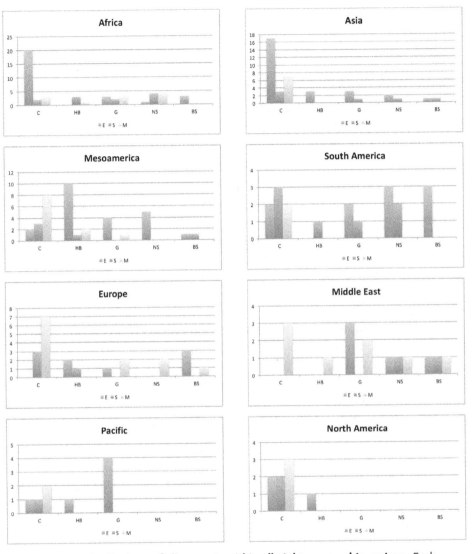

Figure 7. **Distributions of alignments within all eight geographic regions.** Each graph plots the number of equinox (E), solstice (S), and lunar standstill (M) alignments relative to the current geographic pole (C), and the four hypothesized prior pole locations in Hudson Bay (HB), Greenland (G), the Norwegian Sea (NS), and the Bering Sea (BS).

plained alignments of certain megalithic structures at Baalbek, Abydos, Machu Picchu, Ollantaytambo, and in other places. In analyzing the alignment of archaeological sites in Mexico, Aveni and Hartung conclude that an eastern skew was a standard architectural practice over a wide area in Mexico (Aveni, 2001). By accounting for the alignment of all but four of the Mesoamerican sites examined, the shifted pole model explains the reason for the skew.

The geographic pole shift hypothesis also provides a plausible explanation for the apparent lack of astronomical alignments of temples in Upper Egypt (Shaltout & Belmonte, 2005) that is in stark contrast to the precise alignment of numerous pyramids in Lower Egypt to the cardinal directions. As shown in Figure 8, there are more structures in Lower Egypt that are aligned to the current geographic pole than in Upper Egypt. Conversely there are more structures in Upper Egypt that are aligned to previous geographic pole locations than in Lower Egypt. A geographic pole shift from Hudson Bay to the Arctic would have rotated this part of the world approximately 30° (Figure 9) and displaced a considerable amount of water likely inundating low-lying areas along the Mediterranean Sea including Lower Egypt. A sudden shift of the crust would likely have triggered numerous earthquakes along fault lines. The temples in Upper Egypt lie in the Nile River valley far from large open bodies of water and several hundred miles west of the nearest tectonic plate. Perhaps by virtue of their more protected location, structures aligned to previous poles in Upper Egypt survived the crustal displacement and are therefore more numerous than those in Lower Egypt that were likely destroyed at the time.

That a model capable of explaining the alignment of so many archaeological sites that cannot otherwise be explained is itself predicted on Hapgood's unproven hypothesis, is problematic. It is possible that one day Hapgood's hypothesis may be verified by new discoveries in the earth sciences much like Wegner's theory of continental drift was. Although the idea that pole shifts were caused by an asymmetrical buildup of polar ice was rejected at the time, Hapgood's collaborator, J. H. Campbell, developed a model that showed how materials rising out of / sinking into the lithosphere create imbalances in the mass distribution of the crust such that an area of increasing mass has the effect of rotating the crust toward the equator and an area of decreasing mass has the opposite effect of rotating the crust toward the pole (Hapgood, 1958). The Tharsis formation on Mars is an example of how a large mass imbalance is thought to have shifted the Martian poles 20° approximately 3.5 billion years ago (Bouley et al., 2016). Kirschvink et al. (1997)

A) Alignments to current geographic pole B) Alignments to previous geographic poles

Figure 8. Comparison of the number of sites in Lower Egypt (top of each map) and Upper Egypt (bottom of each map) aligned to the current geographic pole A) and previous hypothesized pole locations B).
A) There are 4 sites aligned to the current geographic pole in Upper Egypt and 20 sites in Lower Egypt.
B) There are 3 sites aligned to previous geographic poles in Lower Egypt and 13 sites in Upper Egypt. Photo credit: Google Maps.

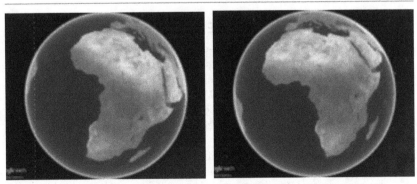

Figure 9. If the North Pole were in Hudson Bay (left), Europe and Africa would be rotated approximately 30° clockwise relative to their current position (right). Photo credit: Google Earth.

hypothesized that the movement of continental landmasses about a half billion years ago shifted Earth's North Pole by 90°.

As noted earlier, Chen at al. (2013) showed that small changes in the weight distribution of the crust caused by climate change induce small changes in the movement of the geographic pole. The current sizes of the Antarctic and Greenland ice sheets are approximately 1.3 x 10^{19} kg and 2.7 x 10^{18} kg, respectively, which are two or more orders of magnitude smaller than Tharsis (10^{21} kg). Twenty thousand years ago the Greenland Ice Sheet is estimated to have been almost ten times larger (Blue Marble, 2017) and could have been much thicker. When the mass of the Greenland Ice Sheet was comparable to that of Tharsis, large changes in it could have resulted in large changes in the movement of the Earth's geographic pole.

In his dismissal of theories of ancient civilizations, Brass (2002) states that there is no paleomagnetic evidence for Earth crustal displacements having occurred. As noted earlier, Kirschvink et al. (1997) concluded from paleomagnetic data collected in Australia and North America that a massive crustal shift did occur between 534 million and 505 million years ago. Paleomagnetic dating methods are intended to measure geological processes that occur over timescales of millions of years. Although it is beyond the scope of the present article to elaborate on this point, the absence of paleomagnetic evidence of Hapgood pole shifts may not be evidence of absence but could be due to the inability of paleomagnetic dating methods to temporally resolve and thus detect climate-induced events occurring over timescales that are two or more orders of magnitude faster than tectonic processes.

REFERENCES

Adhikari, S., & Ivins, E. R. (2016). Climate-driven polar motion: 2003–2015. *Science Advances*, 2(4). Article e1501693. https://doi.org.10.1126/sciadv.1501693

Adhikari, S., Caron, L., Steinberger, B., Reager, J. T., Kjeldsen, K. K., Marzeion, B., Larour, E., & Ivins, E. R. (2018). What drives 20th century polar motion? *Earth and Planetary Science Letters, 502*, 126–132.

Aveni, A. (1980/2001). *Skywatchers of ancient Mexico*. University of Texas Press. pp. 236–238.

Beard, M. (2010). *The Parthenon*. Harvard University Press.

Blue Marble. (2017). Blue Marble: Sea level, ice and vegetation changes—19,000 B.C.–10,000 A.D. https://sos.noaa.gov/datasets/blue-marble-sea-level-ice-and-vegetation-changes-19000bc-10000ad/

Bouley, S., Baratoux, D., Matsuyama, I., Forget, F., Séjourné, A., Turbet, M., & Costard, F. (2016, March). Late Tharsis formation and implications for early Mars. *Nature, 531,* 344–347.

Brass, M. (2002, July–August). Tracing Graham Hancock's shifting cataclysm. *Skeptical Inquirer, 26*(4), 46–49.

Carlotto, M. J. (2019, April 15–17). Archaeological dating using a data fusion approach. Paper presentation at *Signal Processing, Sensor/Information Fusion, and Target Recognition XXVIII,* Baltimore MD. SPIE: The International Society for Optics and Photonics. https://doi.org/10.1117/12.2520130

Carlotto, M. J. (2020). An analysis of the alignment of archaeological sites. *Journal of Scientific Exploration, 34*(1), 13–50.

Chen, J. L., Wilson, C. R., Ries, J. C., & Tapley, B. D. (2013). Rapid ice melting drives Earth's pole to the east. *Geophysical Research Letters, 40*(11), 2625–2630. https://doi.org/10.1002/grl.50552

Daradich, A., Huybers, P., Mitrovica, J. X., Chan, N.-H., & Austermann, J. (2017). The influence of true polar wander on glacial inception in North America. *Earth and Planetary Science Letters, 461,* 96–104. https://doi.org/10.1016/j.epsl.2016.12.036

Hannah, R. (2013). Greek temple orientation: The case of the older Parthenon in Athens. *Nexus Network Journal, 15*(3), 423–443. https://doi.org/10.1007/s00004-013-0169-1

Hapgood, C. H. (1958). *Earth's shifting crust: A key to some basic problems of earth science* (Foreword by Albert Einstein). Pantheon.

Holmes, A. (1944). *Principles of physical geology.* Thomas Nelson & Sons.

Kious, W. J., & Tilling, R. I. (1996). *This dynamic Earth: The story of plate tectonics.* U.S. Government Printing Office. https://pubs.usgs.gov/gip/dynamic/dynamic.html

Kirschvink, J. L., Ripperdan, R. L., & Evans, D. A. (1997). Evidence for a large-scale reorganization of Early Cambrian continental masses by inertial interchange true polar wander. *Science, 277*(5325), 541–545. https://doi.org/10.1126/science.277.5325.541

Lohmann, D. (2010). Giant strides towards monumentality—The architecture of the Jupiter Sanctuary in Baalbek / Heliopolis. *Bollettino di Archeologia.* http://www.daniellohmann.net/dox/lohmann_aiac2008.pdf

Milanković, M. (1932). [Numerical trajectory of secular changes of pole's rotation.] Academy of National Sciences Belgrade. http://elibrary.matf.bg.ac.rs/bitstream/handle/123456789/3675/mm35F.pdf?sequence=1

Muller, R. A., & MacDonald, G. J. F. (1997). Glacial cycles and astronomical forcing. *Science, 277*(5323), 215–218. https://doi.org/10.1126/science.277.5323.215

Napier, W. M. (2010). Palaeolithic extinctions and the Taurid Complex. *Monthly Notices of the Royal Astronomical Society, 405,* 1901–1906. https://doi.org/10.1111/j.1365-2966.2010.16579.x

Petrie, W. M. F., & Murray, M. A. (1903). The Osirion at Abydos (Abtu), Egyptian research account—Ninth year. http://ascendingpassage.com/Osirion-at-Abydos.htm

Shaltout, M., & Belmonte, J. A. (2005). On the orientation of ancient Egyptian temples: (1) Upper Egypt and Lower Nubia. *Journal for the History of Astronomy, 36(3)*, 273–298. https://doi.org/10.1177/002182860503600302

Vigato, M. M. (2015). Lost cities of the Mexican Highlands. *Uncharted Ruins: Looking for remnants of the lost civilization* [blog]. http://unchartedruins.blogspot.com/2015/06/lost-cities-of-mexican-highlands.html

Wilson, C., & Flem-Ath, R. (2000). *The Atlantis blueprint: Unlocking the ancient mysteries of a long-lost civilization.* Delacorte.

Woelfli, W., Baltensperger, W., & Nufer, R. (2002). An additional planet as a model for the Pleistocene Ice Age. arXiv:physics/0204004

Woodworth, D., & Gordon, R. G. (2018). Paleolatitude of the Hawaiian hot spot since 48 Ma: Evidence for a Mid-Cenozoic true polar stillstand followed by Late Cenozoic true polar wander coincident with Northern Hemisphere glaciation. *Geophysical Research Letters, 45(21)*, 11632–11,640. https://doi.org/10.1029/2018GL080787

Journal of Scientific Exploration, Vol. 34, No. 2, pp. 233–245, 2020   0892-3310/20

Mind Control of a Distant Electronic Device: A Proof-of-Concept Pre-Registered Study

PATRIZIO TRESSOLDI

Science of Consciousness Research Group (SCRG), Dipartimento di
Psicologia Generale, Università di Padova, Italy
patrizio.tressoldi@unipd.it

LUCIANO PEDERZOLI

EvanLab, Firenze, Italy; SCRG

ELENA PRATI

EvanLab, Firenze, Italy; SCRG

LUCA SEMENZATO

Dipartimento di Psicologia Generale, Università di Padova, Italy

Submitted May 30 19, 2019; Accepted January 22, 2020; Published June 30, 2020
https://doi.org/10.31275/2020/1573
Creative Commons License CC-BY-NC

Abstract—This study was aimed at verifying the possibility of mentally influencing from a distance an electronic device based on a True Random Number Generator (TRNG). Thirteen adult participants contributed to 100 trials, each comprised three samples of data each of 15 minutes' duration: one for pre-mental interaction, one for mental interaction, and one for post-mental interaction. For each of these three samples, at the end of each minute, the data sequence generated by the random number generator was analyzed with the Frequency and Runs tests in order to determine if there were any changes in the randomness of the sequence. A further 100 trials of three samples each of the same duration were collected during normal functioning of the device, as a control. The only evidence of an effect of distant mental interaction is an increase of approximately 50%, with respect to control data, of the number of samples within which the pre-determined statistical threshold for the detection of a reduction of the randomness was surpassed in both tests. Although the effect of distant mental interaction is still weak, we believe that the results of this study represent a proof-of-concept for the construction of electronic devices susceptible to distant mental influence.

Keywords: mind–matter interaction; random number generators; mind-controlled devices

INTRODUCTION

In this study, we present the findings of a new distant mind–matter (or PK) interaction study aimed not only as a further contribution to this classical line of research, but mainly as a proof-of-concept for a practical application of this phenomenon.

Distant mental activation of electronic equipment, that being without direct contact or electromagnetic means, seems impossible, but it becomes possible if we consider the ability to mentally alter from a distance the activity of random number generators, for example, the 0 and 1 sequences produced by a True Random Number Generator (TRNG).

Testing the possibility of mentally altering the function of random event generators began in the 1970s with the work of Helmut Schmidt, and later became one of the main lines of research within the Princeton Engineering Anomalies Research (PEAR) laboratory, directed by Robert Jahn and Brenda Dunne (Duggan, 2017; Jahn et al., 2007) employing four different categories of random devices and several distinctive protocols. They show comparable magnitudes of anomalous mean shifts from chance expectation, with similar distribution structures. Although the absolute effect sizes are quite small, of the order of 10^{-4} bits deviation per bit processed, over the huge databases accumulated, the composite effect exceeds 7σ ($p \approx 3.5 \times 10^{-13}$).

Even though a meta-analysis of 380 studies up to 2004, related to this phenomenon, showed a small effect and a large heterogeneity in the studies (Bösch et al., 2006; Radin et al., 2006) and was the object of criticism (Varvoglis & Bancel, 2016; Kugel, 2011), by modifying the interaction procedure and the type of data analysis we believe that it is possible to exploit this small effect for practical applications and, specifically, to activate from a distance an electronic device interfaced with a TRNG.

This device, which we have called MindSwitch2, is described in detail in The Electronic Device section of this paper. In contrast to almost all previous experiments, which required participants to mentally generate an increase in 0 or 1 states and then to compare them to a baseline, we simplified the procedure by asking participants to alter the normal random flow of 0 and 1 toward an excess of either 0 or 1. We thought this procedure more efficient with respect to the classical one since the possible effect of the mental interaction was not bound to a specific influence to the random flow of only the zeros or ones.

In fact, the efficacy of this procedure was confirmed in a study by Tressoldi et al. (2016), and a possible explanation for it was posited in the study by Pederzoli et al. (2017). In an initial pilot experiment, and later in a pre-registered experiment, the participants were asked to alter from a distance the function of a TRNG to reach the threshold level, fixed at ±1.65 z-score with respect to the theoretical average value. The number of mentally altered samples in the confirmatory study was 82.3%, versus 13.7% with no mental interaction.

To verify a reduction in randomness, in this study we applied the Frequency Test and the Runs Test present in the suite of tests provided by the National Institute of Standards and Technology (Bassham et al., 2010), as well as a calculation of the mean of the absolute difference between the ones and zeros in each sample (see Methods section).

The Frequency Test calculates the proportion of ones and zeros in a sequence and determines the probability of the calculated value's deviation from what would be expected if the sequence itself were totally random. The purpose of the Runs Test is to determine whether the number of runs of ones and zeros of various lengths is as expected for a random sequence. In particular, this test determines whether the oscillation between such zeros and ones is too fast or too slow.

The mean of the absolute value of the difference between the zeros and ones of each sample is a rough measure of entropy, indicating the extent of deviation from control conditions. The smaller the mean value, the smaller the absolute value of the difference between the zeros and ones.

The decision to implement these measurements derives from the theory that distant mental interaction may favor order where there is disorder, and therefore be able to reduce the randomness of data by increasing the number of zeros and ones, increasing the sequences of identical values (Burns, 2012), or both.

As a control, for each trial three sets of data were gathered, all for the same duration of time, one before, one during, and one after the mental interaction. In this way it is possible to minimize environmental interferences such as temperature or electromagnetic emissions, even though in normal conditions these factors appear to have no effect on the TRNG's activity.

As a further control of experimental conditions, another 100 trials were recorded mainly on the same days as the experimental data, at least one hour before and after with respect to the latter, each comprising three 15-minute samples of data.

Lacking sufficient information regarding both the ideal interaction duration and the most effective mental interaction strategy to use, we left the participants to decide on the most suitable mental strategy for themselves and to choose the duration of influence as either 5, 10, or 15 minutes.

The main confirmatory hypotheses of this study are that:

a) the samples obtained during distant mental interaction contain a higher number of data that exceed the probability cutoff of the Frequency or Runs tests of non-randomness and/or

b) that the means of the absolute differences between the zeros and ones is greater during the mental interaction than in the pre-interaction and the control phases.

Among the data collected immediately after the mental interaction phase, there is some evidence in the literature suggesting that the effect of mental interaction itself may continue for a certain period of time even after the termination of the voluntary interaction (Stanford & Fox, 1975; Tressoldi et al., 2016). The confirmatory hypothesis is that during the post-influence phase the same effects observed during the voluntary mental influence phase will be obtained.

There are no confirmatory hypotheses regarding the differences between these two conditions.

METHODS

Study Pre-Registration

Before any data were collected, the methods on which this study is based as well as statistical analyses of confirmatory hypotheses were pre-registered at https://osf.io/3g95p and at http://www.koestler-parapsychology.psy.ed.ac.uk/Documents/KPU_Registry_1049.pdf

Participants

Experienced and non-experienced participants were recruited among subjects known to the authors. Only those whose previous experience with this type of experiment was known to the authors were considered as experienced.

The participants were five men (average age 48 years; SD = 15) and eight women (average age 46 years; SD = 13), of whom three were experienced and ten were non-experienced.

As specified in the pre-registration, 100 trials were carried out in blocks of 5. Seven participants chose to contribute with 10 trials each, and the remaining 6 each made 5 trials.

The Electronic Device

The device named MindSwitch2, including its software, is described at https://github.com/tressoldi/MindSwitch so that it can be easily reproduced. In a nutshell, it comprises a single-board Raspberry PI mini-computer, a power bank, a TrueRNG, and a USB stick.

During the study, parameters for analysis of the TrueRNG remained fixed: 100 bits/sec for one minute, for a total of 6,000 bits, collected 15 consecutive times for each of the three phases: pre-mental interaction (PreMI), mental interaction (MI), and post-mental interaction (PostMI).

After each minute, the software analyzed the data using the Frequency Monobit Test and the Runs Test from the National Institute of Standards and Technology (Bassham et al., 2010), and, if the statistical analysis gave a p-value $\leq$ 0.05, a visual and auditory signal was activated (an LED was lit for 5 seconds and a 1-sec acoustic signal was emitted).

The results of each of the 15 measurements were recorded on the USB to be exported and analyzed offline. A copy of the raw data[1] is available at https://figshare.com/articles/MindSwitch/8160269.

These parameters were decided after a series of pilot tasks. Before data collection, the preregistered parameter of 200 bits/sec was changed to 100 bits/sec in order to reduce the Raspberry PI processing time.

Procedure

The dates and times of each trial were agreed upon between the participant on duty and the first author. On the agreed day and time, the first author contacted the participant via Skype. After at least one practice attempt to acquire confidence with the procedure, the formal series of trials began at one or at most two per day (e.g., morning and evening), so as to ensure the participant's best mind–body efficiency. The shortest distance from the MindSwitch was approximately 15 km, the longest distance approximately 4,000 km.

Each trial consisted of three successive 15-minute phases: before (PreMI), during (MI), and after mental interaction (PostMI).

The first author activated MindSwitch2, located at least 5 meters from himself in a room with a constant temperature of about 20 °C and far from any electromagnetic energy sources, including the PC used for the Skype connection. During the mental interaction, the first author, after having given the participant the go-ahead to begin the distant interaction, moved away from the monitor for the entire duration of the session and returned to it after the elapsed time to terminate the session.

All participants were given the following instructions:

> Your task is to influence the output of the flux of o's and 1's generated by the TrueRNG connected with our apparatus [they are shown MindSwitch2], reducing or increasing either the o's or the 1's. If you are able to alter this flux of data up to a given threshold, you will activate a red LED and hear an acoustic signal. Do you prefer to directly look at the MindSwitch2 or not?

In the case of an affirmative response, they were able to see the MindSwitch2 via the Skype camera. When the response was negative, for example if the participant believed it to be a distraction during the interaction, the camera was switched off.

Furthermore, they were asked if they wanted to receive the results after each trial or at the end of their participation. The results were summarized to show them the number of MindSwitch2 activations before, during, and after their mental interaction, and a final evaluation as follows: positive (more activations during the mental interaction with respect to the pre- and post- conditions), negative (fewer activations during the mental interaction with respect to the pre- and post-conditions), or neutral (identical number of activations during the mental interaction and the pre- and post- conditions).

In order to not contribute direct influence on the MindSwitch during this interaction, the experimenter moved out of the place where the MindSwitch was located for the entire duration of the session.

Scoring

As described in the pre-registration, the dependent variables were the number of samples reaching a p-value of ≤0.05 for the Frequency Monobit or Runs Tests, and therefore a minimum value of 0 and maximum of 15 for each trial, as well as the average of the deviations of the absolute differences between the zeros and ones.

RESULTS

The number of trials with the higher number of MindSwitch2 activations in the comparison between MI vs PreMI, PostMI vs PreMI, and MI vs PostMI conditions, detected by the Frequency Test, the Runs Test, and both tests together, are presented in Table 1. These are raw values, but given that the total number of trials is 100, they may also be considered as percentages.

TABLE 1

Number of Trials with a Higher Number of MIndSwitch2 Activations in Comparisons of MI vs PreMI, PostMI vs PreMI, and PostMI vs MI Conditions, Detected by the Frequency Test, the Runs Test, and Both Tests Together

	MI vs Pre-MI	PostMI vs Pre-MI	MI vs Post-MI
Frequency Test	29 − 34 [33 − 28]	34 − 31 [33 − 26]	30 − 41 [29 − 31]
Runs Test	24 − 35 [29 − 36]	30 − 36 [29 − 37]	30 − 30 [34 − 28]
Frequency & Runs Tests	**9 − 3** [4 − 4]	5 − 3 [2 − 4]	**9 − 5** [4 − 2]

MI = mental interaction. Pre-MI = pre-mental interaction. Post-MI = post-mental interaction. [ ] = the same data related to the three samples of the control trials, 2nd vs 1st, 3rd vs 1st, 2nd vs 3rd. Differences from 100 are ties. Bold numbers = the main differences.

Means of the Absolute Differences between Zeros and Ones

The number of samples with a higher mean of the absolute differences between 0s and 1s is presented in Table 2.

TABLE 2
Number of Trials with Higher Mean
of Absolute Differences between Zeros and Ones

MI vs Pre-MI	Post-MI vs Pre-MI	MI vs Post-MI
48 – 51 [50 – 50]	45 – 55 [58 – 42]	49 – 51 [40 – 60]

MI = mental interaction. Pre-MI = pre-mental interaction. Post-MI = post-mental interaction. [ ] = the same data related to the three samples of the control trials.

Comment

With respect to the confirmatory hypotheses, the only dependent variable that seems influenced by the MI is the detection of non-randomness by both the Frequency Test and the Runs Test within the same sample of data (see Figure 1).

In short, in favor of the MI effect we see a difference of 6 trials with respect to the PreMI phase; a difference of 4 with respect to the Post-MI phase, of 5 with respect to the first and second control series, and of 7 with respect to the third control series.

Even if, as described in the pre-registration, these differences can be analyzed from a statistical point of view, we believe that applying inferential statistics to these data is inappropriate in that it is not possible to generalize our results to include other participants and experimenters.

In every case the results of a statistical comparison between the 9% of observed events in MI and the 3% of observed events in PreMI, gives a Z value = 1.78, p = 0.036 (one-tailed); the comparison of the 9% observed events in MI and the 4% observed in the control conditions, gives a Z value = 1.43, p = 0.07 (one-tailed).

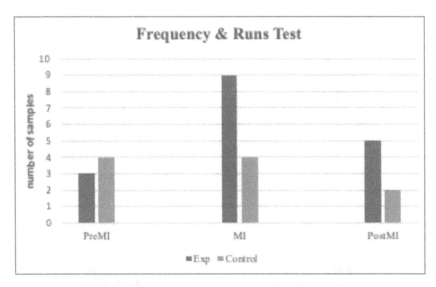

Figure 1. Number of samples where the reduction of randomness was detected by both the Frequency and the Runs tests.

Exploratory Analyses

We wanted to analyze the trend of the absolute differences between zeros and ones recorded in all sample data in the control PreMI, MI, and PostMI phases. Remember that the greater this value, the lower the entropy (randomness) of the sequences of zeros and ones generated by the TRNG.

We therefore counted the number of samples in which these differences exceeded the threshold value of 150, which corresponds to a p-value = 0.05 in the Frequency Test, after which we also did it for those above threshold values of 160, 170, 180, 190, and 200. The results are illustrated in Figure 2.

As shown clearly in Figure 2, the number of samples indicating less entropy, and therefore a larger difference between zeros and ones, is greater in the PostMI condition, followed by PreMI, but the variation disappears when the differences are >190. Furthermore, the number of samples observed in the MI phase is comparable with what is seen in the three control phases.

While for the PostMI phase this result was expected by positing a type of "tail or wave effect" of the MI phase, what was observed in

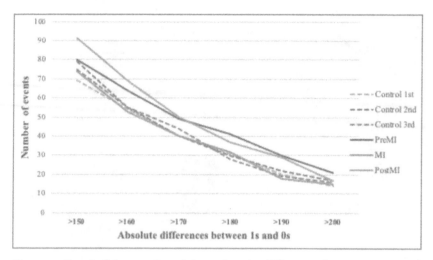

Figure 2. Trend of the number of times that the differences between zeros and ones differed by >150 to >200 in the samples of data of the PreMI, MI, PostMI, and three control phases.

the MI and PreMI phases was unexpected and will be dealt with in the Discussion section.

DISCUSSION

For the time being, mentally influencing MindSwitch2 from a distance does not seem as easy as manually flipping a switch on any type of electronic equipment.

In this experiment the only parameter that appears to be influenced by distant mental interaction is the reduction of randomness detected by the Frequency and Runs tests within the same sample of data. Even though the absolute value is not high—9 samples out of 100—it is however almost twice as much as the PreMI, PostMI, and control phases, as shown in Figure 1.

Do the results of this experiment represent a proof-of-concept of the possibility of creating electronic devices that can be mentally controlled from a distance? We believe so, because our results suggest that it is possible to start to improve the mental-signal/noise ratio of the random number generator ratio.

To reduce the random number generator's noise, apart from seeking those with more stable entropy, new more efficient algorithms to detect reduced randomness could be tested. Furthermore, we still don't know the ideal length of string bits that can maximize the mental signal's effect.

Moreover, how can the mental signal be strengthened? The answer to this question is unfortunately still vague. For example, is there a "dose-effect"—in other words, will the signal improve as the interaction's duration is extended? Of the eight participants who altered the data flow in the MI condition (1) by simultaneously exceeding the statistical thresholds of the two statistical tests, 4 of them had an interaction of 10 or 15 minutes and the other 4 for only 5 minutes. Therefore, this experiment does not seem to highlight a "dose-effect" linked to the duration of the mental interaction.

We wonder if there is evidence of some sort that will allow us to determine whether direct mental influence strategies are more efficient than non-direct mental strategies, such as:

Direct Mental Strategies:

> *I mentally created a flash of light forming a connection cable to MindSwitch.* (Participant #11)

> *I 'asked' and 'hoped' for it to turn on and mentally repeated the request.* (Participant #5)

Indirect Mental Strategies:

> *I attempted, with the aid of spiritual music, to create a field of positive emotion surrounding MindSwitch.* (Participant #8)

> *I cleared my mind of random thoughts.* (Participant #1)

For now, we have no answer to this question either.

Furthermore, if we look at the information in Figure 2, which shows that lower entropy events are more common during the PreMI phase than in the MI phase, still more doubts arise as to the ideal strategy for distant mental influence.

We remind readers that the PreMI phase occurred in the 15 minutes preceding the MI phase, therefore during a time when the participants were certainly not attempting any voluntary influence of

MindSwitch2, but they were indeed preparing to do so by planning the mental strategy to be used after forming a clear image of the end goal.

To summarize, even if we are convinced we have offered a proof-of-concept of the feasibility of a practical application of the mind–matter interaction at a distance with electronic devices, this experiment underscores the many unknowns remaining, before we can improve the "mental signal/noise ratio."

Obviously, these comments are applicable only to what was observed in this experiment. More precise answers will come forth only from further data collection using other participants, other experimenters, other types of random number generators, and analytical algorithms to assess the reduction in entropy of the bits sequences.

NOTE

[1] See the per-participant results in the file "MindSwitch Experiment Summary.xlsx" at https://figshare.com/articles/MindSwitch/8160269

ACKNOWLEDGMENTS

This study was made possible thanks to grant 29/18 from the Bial Foundation and to the voluntary contribution of our 13 participants. We also thank Jim Kennedy for his pre-registration supervision, and the English revision by Cinzia Evangelista in Melbourne, Australia.

REFERENCES

Bassham, L. E., III, Rukhin, A. L., Soto, J., Nechvatal, J. R., Smid, M. E., Barker, E. B., Leigh, S., Levenson, M., Vangel, M., Banks, D. Heckert, A., Dray, J., & Vo, S. (2010). *A statistical test suite for random and pseudorandom number generators for cryptographic applications*. National Institute of Standards and Technology. U.S. Department of Commerce. https://doi.org/10.6028/NIST.SP.800-22r1a

Bösch, H., Steinkamp, F., & Boller, E. (2006). Examining psychokinesis: The interaction of human intention with random number generators—A meta-analysis. *Psychological Bulletin, 132*(4), 497–523. https://doi.org/10.1037/0033-2909.132.4.497

Burns, J. E. (2012). The action of consciousness and the Uncertainty Principle. *Journal of Nonlocality, 1*(1), 1–9.

Duggan, M. (2017). Psychokinesis research. In *Psi Encyclopedia*. Society for Psychical Research. http://tinyurl.com/y6l9zznz

Jahn, R. G., Dunne, B. J., Nelson, R. G., Dobyns, Y. H., & Bradish, G. J. (2007). Correlations of random binary sequences with pre-stated operator intention: A review of a 12-year program. *EXPLORE, 3*(3), 244–253. https://doi.org/10.1016/J.EXPLORE.2007.03.009

Kugel, W. (2011). A faulty PK meta-analysis. *Journal of Scientific Exploration, 25*(1), 47–62. https://www.scientificexploration.org/docs/25/jse_25_1_Kugel.pdf

Pederzoli, L., Giroldini, W., Prati, E., & Tressoldi, P. E. (2017). The physics of mind–matter interaction at a distance. *NeuroQuantology, 15*(3), 114–119. https://doi.org/10.14704/nq.2017.15.3.1063

Radin, D., Nelson, R., Dobyns, Y., & Houtkooper, J. (2006). Reexamining psychokinesis: Comment on Bösch, Steinkamp, and Boller (2006). *Psychological Bulletin, 132*(4), 529–532. https://doi.org/10.1037/0033-2909.132.4.529

Stanford, R. C., & Fox, C. (1975). An effect of release of effort in a psychokinetic task. In J. D. Morris, W. G. Roll, & R. L. Morris (Eds.), *Research in Parapsychology* (pp. 61–63). Scarecrow Press.

Tressoldi, P., Pederzoli, L., Matteoli, M., Prati, E., & Kruth, J. G. (2016). Can our minds emit light at 7300 km distance? A pre-registered confirmatory experiment of mental entanglement with a photomultiplier. *NeuroQuantology, 14*(3). https://doi.org/10.14704/nq.2016.14.3.906

Varvoglis, M. P., & Bancel, P. A. (2016). Micro-psychokinesis: Exceptional or universal? *Journal of Parapsychology, 80*(1), 37–44. https://www.researchgate.net/publication/294675351_Micro-psychokinesis_Exceptional_or_universal

Journal of Scientific Exploration, Vol. 34, No. 2 pp. 246–267, 2020   0892-3310/20

The Global Consciousness Project's Event-Related Responses Look Like Brain EEG Event-Related Potentials

ROGER D. NELSON

Global Consciousness Project
http://global-mind.org

Submitted February 19, 2019; Accepted January 20, 2020; Published June 15, 2020
https://doi.org/10.31275/2020/1475

Abstract—Signal averaging can reveal patterns in noisy data from re-peated-measures experimental designs. A widely known example is mapping brain activity in response to either endogenous or exogenous stimuli such as decisions, visual patterns, or auditory bursts of sound. A common technology is EEG (electroencephalography) or other moni-toring of brain potentials using scalp or embedded electrodes. Evoked potentials (EP) are measured in time-locked synchronization with rep-etitions of the same stimulus. The electrical measure in raw form is ex-tremely noisy, reflecting not only responses to the imposed stimulus but also a large amount of normal, but unrelated activity. In the raw data no structure related to the stimulus is apparent, so the process is repeated many times, yielding multiple epochs that can be averaged. Such "signal averaging" reduces or washes out random fluctuations while structured variation linked to the stimulus builds up over multiple samples. The resulting pattern usually shows a large excursion preceded and followed by smaller deviations with a typical time-course relative to the stimulus.

Keywords: evoked potentials; Global Consciousness Project; time-series, evoked response

The Global Consciousness Project (GCP) maintains a network of random number generators (RNG) running constantly at about 60 locations around the world, sending streams of 200-bit trials generated each second to be archived as parallel random sequences. Standard processing for most analyses computes a network variance measure for each second across the parallel data streams. This is the raw data used to calculate a figure of merit for each formal test of the GCP hypothesis, which predicts non-random structure in data taken during "global events" that engage the attention and emotions of large numbers of people. The data are combined across all seconds of the event to give a representative Z-score, and typically displayed graphically as a cumulative deviation from expectation showing the history of the data sequence.

For the present work, we treat the raw data in the same way measured electrical potentials from the brain are processed to reveal temporal patterns. In both cases the signal-to-noise ratio is very small, requiring signal averaging and smoothing to reveal structure in what otherwise appears to be random data. Applying this model to analyze GCP data from events that show significant departures from expectation, we find patterns that look like those found in evoked potential (EP) work. While this assessment is limited to graphical comparisons, the degree of similarity is striking. It suggests that human brain activity in response to stimuli may be a useful model to guide further research addressing the question whether we can observe manifestations of a world-scale consciousness analogue.

INTRODUCTION

*The surest and best characteristic of a well-founded and exten-
sive induction . . . is when verifications of it spring up, as it were,
spontaneously, into notice, from quarters where they might be
least expected, or even among instances of that very kind which
were at first considered hostile to them. Evidence of this kind is
irresistible, and compels assent with a weight which scarcely any
other possesses.*

—John Herschel (1880/1830)

Since the middle of last century, brain science has been developing sophisticated ways of tapping into neurological activity to learn more about how the brain accomplishes the remarkably complex manifestations of human consciousness. The work is specialized because there are so many kinds of questions, and most answers raise more questions. A major area of research uses measures of electrical potentials as they vary during activity of the brain. One of the most familiar technologies is electroencephalography (EEG) research, with multiple electrodes arrayed over the scalp to capture brain activity corresponding to experiences and activities of the human subject. A sharply focused subset of that technology uses fewer electrodes (an active and reference pair at minimum) to record neural responses from a limited region. Examples are visual evoked responses to a flash of light or an alternating checkerboard pattern, and auditory evoked responses to sound bursts or patterns. The electrical data recording is synchronized to the stimulus onset or pattern, so analysis of the data can identify the onset of the stimulus and track the evoked response over time. Because the data are very noisy, signal averaging is used to compound the data over many epochs. This washes out the unstructured background noise while gradually building up an averaged response to the repeated stimulus. Results are typically presented as a graphical display where variations of the sequential data can be seen in relation to the time of the stimulus.

In this paper we ask a similar question of event-related segments within the database recorded by the GCP over the past two decades. The data are parallel random sequences produced by a world-spanning network of RNGs that record a trial each second comprising 200 random bits. The result is a continuous data history that parallels the history of events in the world over the same 20 years. The GCP was created to ask whether big events that bring large numbers of people to a common focus of thought and emotion might correspond to changes or structure in the random data. Specifically, the hypothesis to be tested states that we will find deviations in random data corresponding to major events in the world. This general hypothesis is instantiated in a series of formal tests applied to events that may engage the attention and emotions of millions of people around the world. For each selected event, analysis parameters including the beginning time, end time, and the statistic to be used are registered before any examination of the data. Over the period from 1998 to 2016, 500 individual tests were accumulated in a formal series whose meta-analysis constitutes the test of the general hypothesis. The bottom line result shows a small but persistent effect with a Z-score averaging about 1/3 of a standard deviation. Though small, the accumulated result over the full database is a 7-sigma departure from expectation, with trillion-to-one odds against it being chance fluctuation. This robust bottom line indicating there is structure in the data supports deeper examination that may illuminate the sources and implications of the anomalies.

Data Characterization

The analysis used for most GCP events is straightforward. For each second, the standardized Z-scores for each RNG in the network are composed as a Stouffer's Z, which is an average across roughly 60 RNGs expressed as a proper Z-score. This is squared, to yield a chi-square with 1 degree of freedom that represents the network variance (net-var) for that second. These are summed across all seconds in the event and normalized to yield a final score. Algebraically, the net-var calculation is closely approximated by the excess pairwise correlation among the RNGs for each second. With 60 or 65 RNGs reporting, there are approximately 2,000 pairs, so this estimate of deviation is robust. Additionally, the pairwise calculation carries more information and allows examination of questions that the simpler measure of composite network variance can't accommodate. For our purposes here, however, the net-var measure is sufficient. We use all the data—the second-by-second scores—representing the longitudinal development during each specified event. In other words, we will be examining the time-series character of the data sequences that define the events.

Data Display

The GCP frequently uses a "cumulative deviation" graph to show the data corresponding to an event selected because it engages mass attention. This type of display was developed for use in process engineering to facilitate detection of small but persistent deviations from the norms specified in manufacturing parameters. It plots the sequence of positive and negative deviations from the expected value as an accumulating sum that shows a positive trend if there are consistent positive deviations, and a negative trend for negative deviations. It looks somewhat like a time series, but because each point includes the previous points, it is autocorrelated (which emphasizes

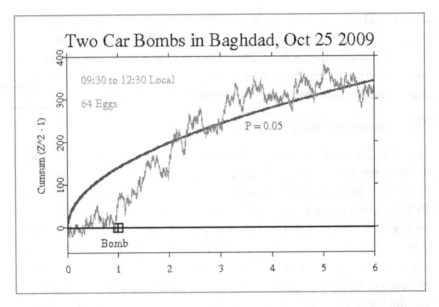

Figure 1. GCP network response to a terrorist bombing in Iraq, October 25, 2009.

persistent departures). Cumulative deviation graphs are well-suited to showing the typically tiny differences from expectation in our data and emphasizing any signal that may be present. The technique mitigates random variation while summing consistent patterns of deviation, thus raising signals out of the noise background.

It will be helpful to look at an example of an event shown graphically in this format. Figure 1 represents the GCP network response to a terrorist bombing in Iraq. It was a global event in the sense that people all around the world were brought to attention and shared emotional reactions. To an unusual degree the thoughts and emotions of millions of people were synchronized. It was a moment in time when we were recruited into a common condition by a major event on the world stage. The event was specified with a duration of 6 hours. This is the most commonly defined event period, which is typically used when something happens that has a well-defined moment of occurrence. The initiating event, in this case a bomb explosion, can be regarded as a "stimulus" to which mass consciousness—and the GCP network—responds.

Reading the graph may benefit from a little instruction. The jagged line is the cumulative deviation of the data sequence, which can be compared against the smooth curve representing the locus of "significant" deviation at the $p = 0.05$ level. The terminal value of the cumulative curve represents the final test statistic, and the curve shows its developing history; it displays, for example, the degree of consistency of the effect over the event period. You can readily see that for much of the period, the data deviations tend to be consistently positive.

Early explorations indicated that any effects we might see in the data take some time, half an hour or more, to develop, followed by two or three hours or more of persisting deviations. Experience brought us to a specification of 6 hours as a period that would usually be long enough to capture any event-correlated deviations, and short enough to distinguish the particular case from the background of ongoing activity in our complex world. It is enough time for most events to affect people local to the event, but also the mass of people around the globe with access to electronic media, radio and television, the Internet, and mobile networks. This example shows a quite steady trend for 3 or 4 hours, after which it levels out, meaning the average deviation is near zero. The endpoint is near the level of statistical significance and the event as a whole contributes positively to the GCP bottom line. It can be thought of as the response of the RNG network during a moment when our hypothesized global consciousness came together in a synchronous reaction to a powerful event.

Though useful, this cumulative deviation presentation obscures the time-course of variations in the raw data, for good cause, as explained above. But our present interest will require starting with raw data to look at structure of a different kind.

Evoked Potentials

An evoked potential (EP) or event-related potential (ERP) is an electrical potential recorded from the nervous system, usually the brain, during and following the presentation of a stimulus. Visual EP are elicited by a flashing light or changing pattern on a computer display; auditory EP are stimulated by a click or tone presented through earphones; somatosensory EP are evoked by electrical stimulation of a peripheral

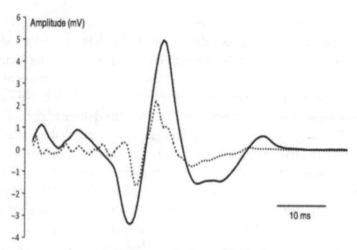

Figure 2. A normal somatosensory evoked potential (EP).

nerve. Such potentials are useful for diagnosis and monitoring in various medical procedures. EP amplitudes tend to be low, and to resolve them against the background of ongoing EEG or other biological signals and ambient noise, signal averaging is required. The recorded signal is time-locked to the stimulus, and, because most of the noise occurs randomly relative to that synchronization point, the noise can largely be canceled by averaging repeated responses to the stimulus.

In Figure 2, positive potentials are up, though graphic displays of EP often use a convention of negative potentials up. This image shows a normal somatosensory EP and is structurally similar to EP in other sensory modalities, with a central peak preceded and followed by a smaller peak with opposite sign. The smooth continuous curve is the result of signal averaging over hundreds of epochs, typically each generated using the same stimulus with locked synchronization of the recording. High frequency noise is reduced by additional smoothing.

Comparison

In the GCP database, each of the 500 formal events can be thought of as analogous to an epoch like those recorded in EP research on human sensory and neurophysiological systems. There is a stimulus in the form of an event that engages the attention of huge numbers of people. It may be a terrorist attack or an earthquake or a mass meditation, but it serves to recruit attention and stimulate synchronous activity in millions of minds. Speculatively, but consistent with the data deviations that correspond to the event, it acts as a stimulus to a global consciousness. This is obviously a model that differs little from poetry—unless we find in the data substantial reason to believe the model is apt and worth exploring. We already have some other indicators that support this kind of model. For example, an examination of the 500 GCP events aggregated in categories such as type of event, size, importance, emotional intensity, and specific emotions such as fear and compassion, shows that "global consciousness" responds much as an individual human does in analogous situations. Another correspondence is that deviations linked with the identified global events are larger when people are awake than at night when they are more likely sleeping. On one level this isn't a big surprise, yet considering that we aren't talking about individual behavior, but an interaction on a global scale, it is thought-provoking.

Yet another indicator of consonance between ordinary human consciousness and hypothesized global consciousness is structure in the event data that is similar in form to what is seen when a sensory stimulus impinges on the human brain. The scale is very different, by a factor of 10,000 or more. The human nervous system typically begins to respond within tens of milliseconds, and the full response to a single visual or auditory stimulus takes half a second or more. Our estimates of GC responses suggest a time period of a few hours. To take a particular example, comparing a half-second brain event to a 3-hour global event gives a ratio of a little over 1 to 20,000. Yet, when we compare responses of these systems with their wildly different scales, we see remarkable similarity in the defining structures.

First, we return to the discussion of raw data versus the cumulative deviation data we ordinarily show in graphical presentations. To process GCP data in a way that is directly analogous to EP data, we must begin with the unprocessed chi-square sequence representing the network-variance response to global events. In Figure 3, the upper left panel shows the usual cumulative deviation plot of data for a composite of nine formal events that showed a significant deviation of the net-var measure. These all are 6-hour events like the example above, but we

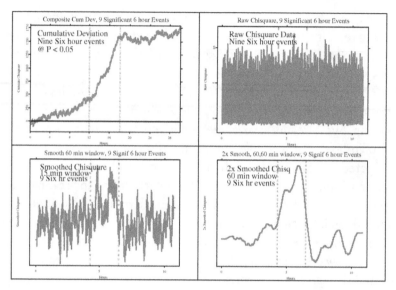

Figure 3. Upper left panel: Cumulative deviation graph for a composite of 9 significant formal events. Upper right panel: Raw data for the composite. Lower panels: Two levels or stages of smoothing the raw data.

are now signal-averaging the events as described for evoked potentials. The other panels in Figure 3 show the raw chi square data and two levels or stages of smoothing, to visualize how the process works.

The data from both research categories, EP and GCP, are noisy and require statistical finesse for analysis (Figure 4). To extract and display signals from the noise background, we use signal or epoch averaging. In brain research, hundreds of measures are taken with data recordings synchronized to the stimulus onset. When these are "stacked" on top of each other and averaged, the random noise tends to cancel and wash out, while any pattern that is linked to the stimulus will gradually build up to show the signal—the time-course of the brain response. Even with a large number of repetitions, the averaged data usually retain high-frequency noise, but this can be mitigated by smoothing. A window encompassing several sequential data points is averaged, then moved to the next point, progressively along the whole sequence. The result is a relatively smooth curve that represents the patterning of amplitude and direction of deviations from the background or baseline activity.

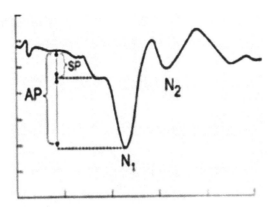

A) Signal-averaged auditory EP

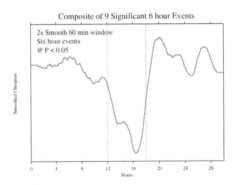

B) Signal-averaged GCP event response

Figure 4. Comparison of an EP graph with a GCP graph.
   A) EP from an auditory stimulus.
   B) Composite of GCP data from nine 6-hour events.

Figure 4A and Figure 4B allow a visual comparison of an EP graph with a GCP graph. The EP example, Figure 4A, shows the evoked potential from an auditory stimulus. It is an example of data gathered in clinical research (Anbarasi, 2019). Figure 4A is described as a normal electrocochleaogram (OCoG), and it displays signal-averaged data from electrodes placed trans-tympanically into the cochlea. It uses the convention found in much of the evoked potential literature showing negative potentials upward. It is typical in displaying a large primary spike with smaller variations before and after, some of which are sufficiently distinct and regular as to be labeled.

Figure 4B is an example of GCP data treated in the same way. This is a signal-averaged composite of data from nine of the 6-hour events described earlier. These were chosen because they show a clear effect as indicated by a significant terminal deviation. The whole dataset for each event includes 12 hours before and after the event period, for a total of 30 hours. As described earlier and shown in the four-panel Figure 3, we use the raw data (net-var measure at 1 per second) rather than the cumulative deviation of the net-var, in order to parallel what is done in EP research. (You may recognize Figure 4B as an inverted version of the lower right panel of Figure 3.) Following the analysis procedures for EP, the signal-averaged raw GCP data are smoothed with a moving (sliding) window long enough to reveal the major structure. For the 6-hour events, an appropriate window is 3,600 seconds. High-frequency noise is then mitigated by a second pass. The result is a smooth curve representing the major (low band-pass) variations of the data during the events. The structure represents the common features across repeated measures of data deviations during major events.

The signal-averaging process was also applied to a sample of 24-hour events in the GCP database (Figure 5). There are 12 such events meeting the significance criterion, making them likely cases of a real effect correlated with the specified events. The 24-hour event data are surrounded on both sides by 24 hours of non-event data. The same kind of smoothing with a coarse and fine pass was used as for the 6-hour events, so the smooth curve represents a low band-pass filtering of the raw data. For the EP comparison, we show a positive up-trace of an auditory evoked potential.

The visual matching in this case is as compelling as the 6-hour event example, but the variability of data in both domains is large even with statistical smoothing. EP research shows a wide variety of detailed graph shapes, but there is a common theme: small shifts in one direction, followed by a larger, primary shift in the opposite direction, then a return to baseline and often a small opposite peak or damping oscillation.

Interpretation

Many interesting questions are brought into view by the comparison of EP versus net-var structure. There are differences, of course, beyond those relating to scale and to physical versus statistical measurement.

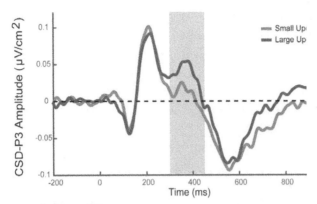

A) Signal-averaged sensory EP

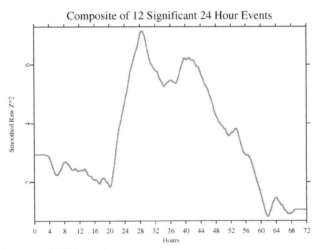

B) Signal-averaged GCP event response

**Figure 5. Comparison of A) positive up-trace auditory EP, with
        B) GCP composite of 12 24-hr events.**

Yet it is worthwhile to think further about some of the questions.

It seems important, given the fundamental character of the EP model, to consider what constitutes the "stimulus" to which the subsequent response is linked. In EP research that's unambiguous—it is literally imposed by the experimental technology. In the GCP case, the stimulus isn't quite so clear, though we can make a case that, at least for the 6-hour events, it is the point event to which the world responds. That, by definition, occurs near the beginning of the event. But, is there a post-stimulus delay—the equivalent of the 10 to 50 ms in EP measures between the stimulus and the first big spike in voltage? In the examples shown here, such a delay isn't easy to identify, though there is some structure that might qualify.

The GCP epochs averaged in the first comparison are 6 hours in duration, surrounded by 12 hours preceding and following the formal event, with the "stimulus" roughly at the beginning of the event period. The stimulus in the 24-hour figure might be posited at the 24-hour point marked by the vertical line, but in most of these cases the effective stimulus is episodic or distributed over the 24-hour period.

There are speculative suggestions worth considering. Many events in the GCP experiment are in a strong sense internally defined. That is, the event exists only when it happens, so it is its own stimulus. This is most obviously the case for 24-hour events such as organized meditations and demonstrations. It may also be of value to think of endogenous stimuli. For example, a decision to act, say move a finger, may appear in the EP data before it appears in consciousness. We note that the 24-hour subset does show a building response before the event period begins. A moving average incorporates later data into the present calculated point, but only about 30 minutes of the apparent 3–4-hour early buildup can be attributed to the mathematical smoothing process.

The primary research question is how any stimulus translates into a structured response in the random data from the GCP network. Why do our physical random devices become correlated at times when the thoughts and emotions of many humans become synchronized and coherent? The data say this is no accident or coincidence, and the experimental design ensures these correlations are meaningful. Could that widespread coherence generate an information field with the capacity to produce correlations in the random data? Do the intentions and expectations of researchers enter into the definition and execution of an experiment with results showing structure in what should be random data? There are multiple "explanations" for the small but highly significant data deviations, but thus far none is fully satisfying. Probably we need to look for explanations that recognize and integrate multiple sources.

Global Consciousness versus Goal Orientation

It seems appropriate to look at the material that stimulated this excursion into analogues for the GCP event data. Peter Bancel spent many years doing careful post hoc analysis on the GCP database looking for information and parameters to define a global consciousness (GC) model. He worked progressively toward demonstrations that generalized field models were a good fit to the data, and showed they were significantly better than another major contender, DAT-like selection models that posit precognitive information about future results driving present choices (e.g., when to start the experiment) (Bancel & Nelson, 2008; Bancel, 2011; Nelson & Bancel, 2011). His most direct presentation of the case for field-like models was a 2013 paper submitted for presentation to the Parapsychological Association annual meeting (Bancel, 2013). Not long thereafter, Bancel reversed his position and began describing and promoting a goal orientation model (GO) that is essentially the DAT approach he had earlier rejected (Bancel, 2015).

The GO model postulates psi-based experimenter selection of parameters, in particular the starting and ending points of the events. This model addresses only the primary measure, and is incapable of dealing with other structural elements of the GCP data, but Peter argues that GC can't work, for technical and philosophical reasons. He supports his argument by a graphical analysis, shown in Figure 6A. It is from a paper summarizing Peter's views on the most suitable model for GCP findings (Bancel, 2017).

The Figure 6A graph shows reversals at event boundaries that justify a preference for GO by conforming to an idealized selection model. Figure 6A is a composite of all short GCP events, which nominally allow the experimenter to select start/end times. (This is in fact not the case for a large proporton of the events. For example, many events are repetitions that use the prior specifications, or use timing drawn from media reports.) The proposal is that experimenter psi can achieve a desired future result by selecting from the naturally varying data sequence an appropriate deviant segment. Further, Bancel argues that selecting points in the data sequence that define a positive segment will cause the preceding and following segments to show a deficit or a negative tendency (Personal communication, July 8, 2016):

> If there is a choice of how to partition a null dataset, so that the chosen segment has a mean >0, then the remaining segment will necessarily (on average) have a mean <0. Choosing a start time is like this because the choices all are relatively proximate: You realistically might choose a time a minute earlier or later; or 15 minutes earlier or later; but not 12 hours or 12 days earlier or later.

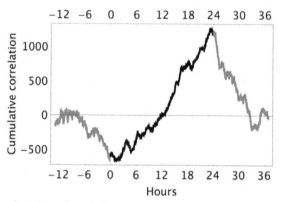

A) Cumulative deviation, short GCP events

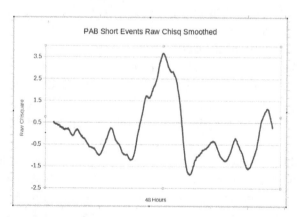

B) Smoothed raw data, short GCP events

Figure 6. A) Cumulative deviation, short GCP events (from Bancel, 2017).
 B) Smoothed raw data, short GCP events (derived from Figure 6A).

I think this argument is fallacious—not least because it sounds like the gambler's fallacy (Bennet, 2019), given that the "null dataset" is by definition random and is continuous over years. The "balancing" seen in the composite figure clearly needs a better explanation.

Something about this graphical presentation tugged at my unconscious for months—rooting around in old memories looking for images akin to this oscillating picture. Finally, it bubbled up to the surface. The graphic was reminiscent of event-related neurophysiological measures, which also show an oscillating response, albeit with a different shape. To see the connection more clearly it was necessary to revert to raw data, as described earlier. In order to process these data using the EP procedures, I decomposed Bancel's original cumulative deviation figure to produce a file of equivalent raw data and proceeded with smoothing. The result is shown in Figure 6B. It bears out my intuition that it should look like EP data.

The cumulative deviation graph of the GCP "short" events shows sharply delineated inflections at the event boundaries, even though it includes a large proportion of null and negative outcome events, and still more events with previously determined, fixed parameters (there is no selection). The precision of the fit to the idealized model is surprising, given the large proportion of events that do not conform to the required conditions. Perhaps the shape of the curve has another source than the proposed, goal-oriented psi data selection. The smoothed raw data graph, mimicking physiological EP measures, suggests a viable candidate.

Bancel made a similar figure for all the GCP formal data, first normalizing all the various event lengths to a 24-hour standard (Figure 7A). A context of 24 hours before and after was included in the plot, and as in the case of the short event example, there are inflections at the event boundaries, and negative-going trends before and after. He argues that this supports the GO psi-selection model, but, as in the previous case, there are many exceptions—events that explicitly do not conform to the required model criteria where selection is allowed. And again, there is an alternative "explanation" for the shape of the curve, namely an event-related potential model. The graph of smoothed raw data, Figure 7B, derived from the "all events" figure is practically indistinguishable from typical EP graphs.

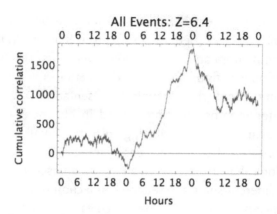

A) Cumulative deviation, all GCP events (normalized to 24 hrs)

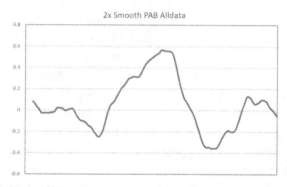

B) Smoothed raw data, all GCP events (derived from Fig 7A)

Figure 7. A) 72-hour context, all GCP events (from Bancel, 2017).
            B) Smoothed raw data, all GCP events (derived from Figure 7A).

An Independent Look

Dean Radin in the course of his peer review of this paper (personal communication, October 16, 2019) performed simulations that directly compared the two models and found no support for the GO perspective:

> I haven't done any more simulations recently, but from what I did look at I see why positive trends would appear before and after an event. That's due to the dependencies introduced by smoothing. But I don't see how those trends would

end up being negative. That doesn't make sense logically nor is it what the simulations show. . . . [M]y sense is that Peter's [Bancel]argument doesn't stand up.

These observations support my contention that some other explanation is needed for the shape of the cumulative deviation curves than that proposed by Bancel. His assertion that a selection model would produce negative deviations before and after the positive trend of the event data segment is not only logically dubious but is specifically not supported in appropriate simulations.

A Single Event

While the comparisons described above depend on signal averaging across multiple events meeting a criterion of significance, we can ask if a sufficiently powerful individual event might show the same kind of structure. One that stands out in the GCP database is the terrorist attacks on September 11, 2001. The GCP network had been in place for three years and the number of Eggs (Electrogaiagram) had grown to 37, so the data recorded on September 11 were statistically robust. Because it was such a clear instance of an event that should instantiate the GCP hypothesis, we paid close attention. In addition to the a priori specified hypothesis test, we looked at other aspects of the event and did other analyses, including one extending the time period to include a context of 9 days around the event. The standard net-var calculation was applied to data from September 7th to September 15th. The slope of the cumulative deviation graph beginning when the first World Trade Center tower was hit and continuing for nearly three days is extreme. An informal estimate for the probability lies between 0.003 and 0.0003 (this means an odds ratio on the order of 1 in 1,000). Visual inspection (Figure 8) suggests the trend begins as much as a day before the planes crashed into the World Trade Towers and continues for more than two days after the towers fell.

Though the time-scale differs, the cumulative deviation graph for this singular event presents a picture that is much like that seen for the signal-averaged events shown above, leading us to ask what structure the corresponding raw data might show when processed using the EP protocol and low band-pass filtering.

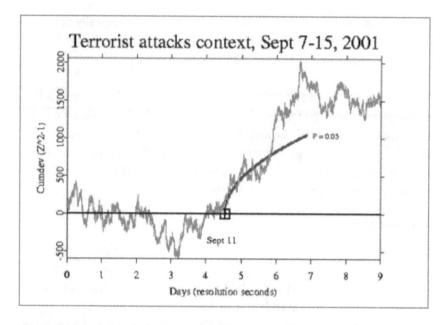

Figure 8. Cumulative deviation graph of the September 11, 2001, terrorist event.

In Figure 9A, we see an answer to that question. The graph of smoothed raw data from the 9/11 context analysis does look like EP data, as can be seen here. It has the general form we have seen before, with a large deviation bracketed by smaller deviations of opposite sign. For a comparison, Figure 9B shows an example of evoked potentials recorded during tests of four cognitive processes: action-effect binding, stimulus-response linkage, action–effect feedback control, and effect–action retrieval. While I chose this picture because it is a good match, it is representative of a broad class of event-related potentials.

DISCUSSION

We have seen multiple examples of striking similarity between event-related brain potentials and event-related correlations in random data. Is the GCP network of widely distributed random number generators picking up something like the evoked responses of an earth-scale consciousness to powerful stimuli? If that idea is to be given serious consideration, how can the timing of the 9/11 "response" be explained? It can't be regarded as an immediate response to the terrorist attacks

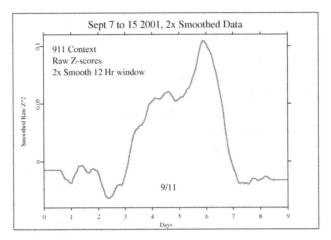

A) Smoothed raw data for 9 days around 9/11

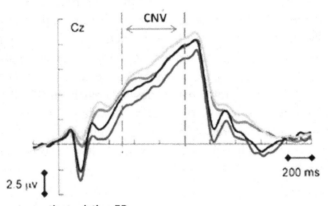

B) Contingent negative variation EP

**Figure 9. A) Smoothed raw data from 9/11 context analysis looks like EP data.
B) Brain evoked potentials during tests of four cognitive processes.**

because the apparent changes begin more than a day earlier. Could the small group of 50 or 100 terrorists planning and working toward the attack be responsible? That would be counter to the experience and findings of the Field REG studies of group consciousness. And it would be inconsistent with findings in the GCP database, where coherence among small numbers of people is associated with small effects. It is arguably just as likely that a global consciousness, whatever its nature, might manifest presentiments of the future, given an emotionally powerful stimulus, just as humans do (Radin, 2004). We can even calculate roughly the dimension of the former. The ratio of global-scale response times to the time-course of human perception is on the order of 20,000 to 1 (Nelson, 2019). The presentiment response shows up in physiological data on the order of 3 to 10 seconds before the stimulus. This corresponds in the GCP data to 0.7 to 2.3 days—in the same ballpark with the examples presented here.

These analyses are interesting on multiple levels, and they raise good questions. It is premature to claim that the visual comparisons make a rigorous case akin to direct measures like recordings from the brain in EEG and EP work. We have only correlations and concordance. On the other hand, the conformance of event-related GCP responses to the general patterns of stimulus-related brain potentials is noteworthy. All the examples we have seen support the idea that the GCP network reacts to the stimulus of global events with temporal variation that practically duplicates the response of neural networks to relevant sensory stimuli. This explanation for the shape of the GCP data curves is arguably better than the experimenter psi-selection model proposed by Bancel. It is considerably more "down to earth" in that it requires no precognition of future system states to guide present choices. And there is no conundrum regarding events with fixed parameters or null and negative results. It is comfortably compatible with some temporally local, field-like model. While we can't formally describe a mechanism that can connect a mass consciousness response to the RNG network deviations, there is a clear, well-established correlation. Notably, if we take a serious look, that is all we have in the evoked potential case as well—just established correlations. Yet, neurophysiologists use EPs for diagnosis and treatment with no further ado.

Almost all psi research depends on statistical rather than direct measures. But it can be argued that correlation is a thing, "*ein Ding an sich,*" and it is worth some effort to flesh out that proposition (Atmanspacher, 2018). Can we draw an equivalence between statistical and physical measures? It is, at base, the same question as the more general one about information. Is it possible to formulate a relationship of information and energy that is like the one established early in the last century for energy and matter? If that happens, it will clarify important issues, not only in psi research but more broadly in science and philosophy.

We will need a lot more data and much deeper thought to resolve such questions.

REFERENCES

Anbarasi, M. (2019). Auditory evoked potentials in clinical research. https://www. slideshare.net/anbarasirajkumar/evoked-potential-an-overview?next_slideshow=1

Atmanspacher, H. (2018). Synchronicity and the experience of psychophysical correlations. In Christian Roesler (Ed.), *Research in analytical psychology: Empirical research* (pp. 227–243). Routledge.

Bancel, P. (2011). Reply to May and Spottiswoode's 'The Global Consciousness Project: Identifying the source of psi'. *Journal of Scientific Exploration, 25*(4), 690–694. https://www.scientificexploration.org/docs/25/jse_25_4_Bancel.pdf

Bancel, P. A. (2013, August 8–11). Is the Global Consciousness Project an ESP experiment? Submitted for presentation to the 56th Annaul Parapsychological Association Convention. Institut Métapsychique International, Paris, France.

Bancel, P. A. (2014). An analysis of the Global Consciousness Project. In D. Broderick, & B. Goertzel (Eds.), *Evidence for psi: Thirteen empirical research reports.* McFarland.

Bancel, P. A. (2017). Determining that the GCP is a goal-oriented effect: A short history. *Journal of Nonlocality, 5*(1). https://journals.sfu.ca/jnonlocality/index. php/jnonlocality/article/download/70/70

Bancel, P., & Nelson, R. (2008). The GCP event experiment: Design, analytical methods, results. *Journal of Scientific Exploration, 22*(3), 309–333. https:// www.scientificexploration.org/docs/22/jse_22_3_bancel.pdf

Bennett, B. (2019). Gambler's fallacy. Logically fallacious. https://www. logicallyfallacious.com/tools/lp/Bo/LogicalFallacies/98/Gamblers-Fallacy

Herschel, J. F. W. (1880/1830). *Preliminary discourse on the study of natural philosophy.* Longman, Rees, Orme, Brown & Green. [Original published 1830]

Nelson, R. D. (2019). *Connected: The emergence of global consciousness.* ICRL Press.

Nelson, R. D., & Bancel, P. A. (2011). Effects of mass consciousness: Changes in random data during global events. *EXPLORE, The Journal of Science and Healing, 7*(6), 373–383. https://doi.org/10.1016/j.explore.2011.08.003

Radin, D. I. (2004). Electrodermal presentiments of future emotions. *Journal of Scientific Exploration, 18*(2), 253–273. http://deanradin.com/articles/2004%20 presentiments.pdf

Journal of Scientific Exploration, Vol. 34, No. 2, pp. 268–350, 2020   0892-3310/20

Behind The Mask:
Decoding the Dedication of Shakespeare's Sonnets

PETER A. STURROCK
Applied Physics Department, Stanford University, Stanford, CA

KATHLEEN E. ERICKSON
School of Information, San José State University, San José, CA

Submitted September 18, 2019; Accepted January 20, 2020; Published June 15, 2020

https://doi.org/10.31275/2020/1672
Creative Commons License CC-BY-NC

A shorter version of this paper was given by Peter Sturrock as an invited Founder's Lecture on June 7, 2019, at the 38th Annual Meeting of the Society for Scientific Exploration in Broomfield, Colorado.

Abstract—The Dedication of Shakespeare's Sonnets has long been a mystery that orthodox (Stratfordian) scholars have been unable to resolve, the reason being that the true message is not evident—it is concealed in cryptograms. We here address the Authorship Issue (Who was the true author of the monumental literature we attribute to "Shakespeare"?) from a scientific perspective. We follow the initiatives of John Rollett, Jonathan Bond, and David Roper, who brought their mathematical expertise to the challenge of identifying and deciphering cryptograms embodied in the Dedication of the Sonnets and in the Inscription on the "Shakespeare" Monument. We show that the combined statistical significance of the cryptograms is overwhelming, so that the messages must be accepted as the intended creations of the authors—Edward de Vere for the Dedication and Ben Jonson for the Inscription. The cryptograms confirm that *Shakespeare* was the mask adopted by de Vere, 17th Earl of Oxford, as proposed by J. Thomas Looney in 1920, and that the intended recipient of the Sonnets was Henry Wriothesley, 3rd Earl of Southampton. In combination, the cryptograms denote a loving, possibly intimate, relationship between de Vere and Wriothesley.

Keywords: Shakespeare authorship; cryptograms; Edward de Vere; 17th Earl of Oxford; Shakespeare Dedication; Shakespeare Monument; William Shakspere; Henry Wriothesley

Everything seemed to point to his being but a mask, behind which some great genius, for inscrutable reasons, had elected to work out his own destiny.

—J. Thomas Looney (1920, p. 3)

1. INTRODUCTION

The plays, Sonnets, and other poems we attribute to William Shakespeare (or Shake-Speare) are widely and justifiably recognized as the greatest contribution to the literature of the English language.

This being the case, one would imagine that all scholars who have an interest in the work of Shakespeare would wish to know as much as possible about his identity: What was there about his parentage, schooling, and life experiences that can begin to explain his knowledge of the world—his highly detailed knowledge of France and Italy (including their languages), his knowledge of English history and court life (including court protocol and the pastimes of the nobility), his knowledge of botany, medicine, and many other fields (especially his highly detailed and accurate knowledge of the law), his knowledge of the classics (especially his familiarity with the works of Ovid), etc., etc.?

Scholars have no persuasive answers to any of these questions since the orthodox doctrine identifies Shakespeare the great author with a man who was baptized as, and typically used the name of, William Shakspere, born and raised in the small town of Stratford-upon-Avon in the West-Midlands county of Warwickshire. The usual suggestion of an answer is "He was a genius." But the greatest genius can process and build upon only the information he ("or she" understood throughout as appropriate) has acquired and assimilated as part of his life experiences.

We have some understanding of the origin of this doctrine, but we have no understanding of its persistence, except to note that—as Shakespeare wrote—one can sometimes become "tongue-tied by authority." If this is so, progress may require the efforts of one or more scholars who are not subject to "authority"—more specifically, scholars who are not members of the English-Literature Establishment—for instance, mathematicians or engineers.

How could mathematicians possibly contribute to the resolution of a question of literature? This is, admittedly, an unlikely event—unless the literary problem happens to involve cryptograms, in which

Figure 1. J. Thomas Looney

case a mathematician has a big advantage over any non-mathematician. This claim is the subject of this article.

Even a non-mathematician can make important progress if he thinks along scientific lines. So it was with J. Thomas Looney (see Figure 1), who initiated the current insurrection against the orthodox doctrine in 1920 with the publication of *"Shakepeare" Identified in Edward de Vere, the Seventeenth Earl of Oxford* (Looney, 1920). Although not a scientist (he was a schoolteacher), Looney proceeded in a way that any scientist would recognize and appreciate: *He began by identifying and then reviewing the relevant facts.* This is the crucial distinction between the work to be described in this article and the work of Establishment scholars who typically try to fit the facts to the received theory.

The current orthodox doctrine is based on the *assumption* or the *theory* that William Shakespeare, the great author, was William Shakspere, an otherwise unremarkable—and possibly illiterate—person baptized on April 26, 1564, in Stratford-upon-Avon. Orthodox scholars then face the challenge of reconciling the few facts we have about Shakspere with the extraordinary—and so far unequalled—literary output that we attribute to Shakespeare. Scholars have attempted to make this problem somewhat more tractable—or to appear somewhat more tractable—by replacing the actual name of *William Shakspere*, or variants thereof, with the name *William Shakespeare*, which Shakspere never used.

Looney's great contribution was to show that a careful analysis of the facts leads to the conclusion that "William Shakespeare" was not the name of a resident of Stratford-upon-Avon or of London, and was not the name of any known poet or playwright, but the *nom de plume* adopted by a nobleman, Edward de Vere, 17th Earl of Oxford.

The suspicion that "William Shakespeare" might be a *nom de plume* has a long history. Many names have been suggested for the identity of the great author. In the early 20th Century, a prime candidate for authorship was the erudite Sir Francis Bacon, the author of memorable but somewhat ponderous prose. (Think of *"'What is Truth?' asked Jesting Pilate, and would not stay for an answer . . . "*)

The case for Sir Francis Bacon was advocated in the early 19th Century by Delia Bacon, an American woman who, she pointed out, was unrelated to Sir Francis. In 1856, she published an article in *Putnam's Monthly* on "Shakespeare and His Plays: An Inquiry Concerning Them" (Bacon, 1856). She followed this up in 1857 with a 543-page volume entitled *The Philosophy of the Plays of Shakespeare Unfolded.* Elizabeth Wells Gallup, also an American woman, also spent years searching for cryptograms in the Shakespeare plays (Gallup, 1910). Delia Bacon and Elizabeth Wells Gallup both claimed to find evidence for Sir Francis secreted in some of the Shakespeare plays.

It appears that the Folger Shakespeare Library sought the opinion of two professional cryptographers, William F. and Elizebeth S. Friedman, who were world-renowned for their critical role in breaking Japanese codes in the tense years leading up to Pearl Harbor. The Friedmans carried out a highly detailed analysis of the Bacon–Gallup proposals for cryptographic content of the Shakepeare oeuvre, and concluded that they could find *no evidence of hidden messages such as had been proposed by Delia Bacon* (Friedman & Friedman, 1957). However, the Friedmans—presumably following the Folger initiative—restricted their attention to the type of cryptogram used by Delia Bacon—the *biliteral cipher.* Had the Friedmans carried out a more general investigation, they might have discovered cryptograms of a type not envisaged by Delia Bacon. The Friedmans subsequently received an award from the Folger Library.

The next serious investigation of possible cryptograms in the works of Shakespeare was carried out not by an academic Shakespeare scholar, nor by a professional cryptographer, but by an electrical engineer. John M. Rollett discovered three cryptograms in the Dedication of Shakespeare's Sonnets. Rollett (who passed away in 2015) and his discoveries are the subjects of Sections 5 and 6.

It is relevant to note that Rollett, as an engineer responsible for advanced projects in the main telecommunications laboratory in Britain, had a more-than-adequate knowledge of the kind of mathematics necessary for determining the significance—or insignificance—of any patterns one might find secreted in apparently innocent text.

Later contributions by Jonathan Bond and David Roper will be discussed in the Sections 7, 8, 9, and 10. (Bond, Roper, and Sturrock were all trained as mathematicians. Bond and Roper are also Latin scholars.)

The independent scholar Diana Price has carried out research on the life of William Shakspere (Price, 2012). As part of this research, Price has drawn up a chart that compares what is known of Shakspere with what is known of 24 writers in England whose lives overlapped with the life of Shakspere. It proves possible to analyze this chart mathematically in order to evaluate the probability that Shakspere was a writer like the 24 comparison authors (Sturrock, 2008). This analysis is discussed briefly in Section 4.

The work of Bond, Looney, Price, Rollett, and Roper has been in the open literature for decades, yet it is still possible for a student to spend six to nine years at a major university in Britain or the United States, studying English literature and acquiring a BA, an MA, and a PhD along the way, and not even learn that there is a significant Shakespeare Authorship Question. (In some universities they might only learn that an American lady named Delia Bacon (1856) had the unsubstantiated idea that the works of Shakespeare were written by Sir Francis Bacon, and that she died in an asylum.)

Why do we care? Why *should* we care? Not everyone does care. We have heard a good friend remark *Why does it matter who wrote the plays? We have the text—knowing the name of the author is not going to change the text!*

To which we reply—*When we listen to Beethoven, we also think of Beethoven. When we read* The Life of Samuel Johnson, *we think of Samuel Johnson and James Boswell. When we look at a Picasso, we think of Picasso. Our perception of the music or text or painting is influenced by our knowledge of—and our feelings for—the composer or the writer or the artist. There is no real separation. What we hear or read or see informs our knowledge of—and our appreciation of—the man and his life and the event of this creation—and vice versa.*

Suppose that, in all the libraries and conservatories of the world, all references to *Ludwig Van Beethoven* were removed and replaced by the name *Josef Schmidt*, a man who could not even play the fiddle or whistle a tune. Would we not consider that not only a dereliction of scholarship but also a catastrophic injustice?

What would be the difference between erasing the identity of the great composer we know as *Beethoven*, and erasing the true identity of the great poet and playwright we know as *Shakespeare*?

Some scholars do care about the potential injustice—and dereliction of scholarship—of possibly attributing the poems and plays of Shakespeare to the wrong person. Regrettably, they tend not to be taken seriously.

Furthermore, there is often—perhaps typically—a subplot, or hidden agenda, to Shakespeare plays, as has been explained in some detail by Eva Lee Turner Clark (Clark, 1931).

The conventional attribution of the authorship to William Shakspere of Stratford-upon-Avon has become a doctrine that it is inexpedient and unwise to question. Resistance to the study of the Shakespeare Authorship Question seems to be more a political issue than a scholastic one.

We discuss some of the basic facts about Willliam Shakspere and Edward de Vere in Sections 2 and 3, respectively. More of their life events are noted in Table 1, which is located at the end of this article.

2. WILLIAM SHAKSPERE—THE ORTHODOX CANDIDATE

According to the orthodox "Stratfordian" doctrine, the great author whom we know as *Shakespeare* was born, lived much of his life, and died and was buried in the small town of Stratford-upon-Avon in the county of Warwickshire in the West of England.

What records do we have of such a man? None—but we do have a few records of someone with a similar, but not identical, name.

A man who went by the name of *William Shakspere* or variants thereof (all with a short "a" as in "cat", not a long "a" as in "bake") was born in Stratford-upon-Avon in 1564. His baptismal record, dated 26 April 1564, reads *Guilielmus filius Johannes Shakspere*. His burial record, dated 25 April 1616, reads *Will. Shakspere gent*.

On November 27, 1582, a certificate issued at the nearby city of Worcester provided for *William Shaxper* to marry *Anne Whateley* of Temple Grafton. Whether a man of that name actually married a lady of that name, we do not know, and is the subject of some intriguing speculation.

However, we do know that the very next day (November 28, 1582), a certificate was issued in Worcester that gave *William Shagspere* permission to marry *Anne Hathaway* of Shottery, and that this marriage did take place, *Anne Hathaway* becoming *Anne Shakspere*. At the time of their marriage, William was eighteen years old and Anne was twenty-six. Their first child, *Susanna Shakspere*, was baptised on May 26, 1583, according to the Holy Trinity Church parish register. Their next children were twins, baptized as *Hamnet Shakspere* and *Judith Shakspere* on February 2, 1584 (named after neighbors, see Table 1).

Scholars have found a few legal records—related to non-payment of taxes, purchases of grain, suits to recover unpaid loans, etc.—all in the name *Shakspere* or a similar version with the short "a." Shakspere was a successful businessman who acquired considerable property and was one of the wealthest citizens of Stratford-upon-Avon when he died. The salient known facts about Shakspere's life are listed, by date, in the Table 1.

There are no legal records that tie *William Shakspere* to any literary or related activities, as we shall see in Section 4. There are in fact reasons to suspect that William Shakspere was illiterate—which was the norm rather than the exception for low or middle-class citizens in England in the Sixteenth Century.

It is surely significant that the death of Shakspere was a non-event (no eulogy, no state funeral, no move to have him buried in Westminster Abbey). By comparison, the playwright Francis Beaumont (who died in 1616, the same year as Shakspere) was buried in Westminster Abbey.

The six signatures that scholars attribute to "Shakespeare" are shown in Figure 2. The first signature in Figure 2, dated May 11, 1612, was on a deposition in what is known as the "Mountjoy case." Shakspere was called to be a witness in a case concerning a dowry that was promised and (according to the petitioner) reneged on. (Shakspere was said to have been the broker of the marriage transaction, but he claimed to have no recollection of the agreement.) Signatures 2 and 3, dated March 11, 1613, appear on documents related to the purchase of the "Blackfriars Gatehouse." Signatures 4, 5, and 6 all appear on

Signature on the Mountjoy Deposition, May 11, 1612

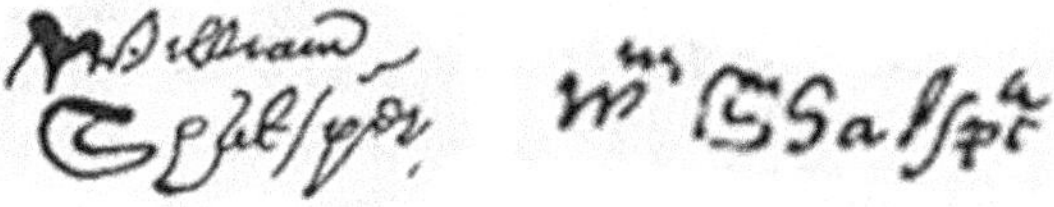

Signatures on the Blackfriars Documents, March 11, 1613

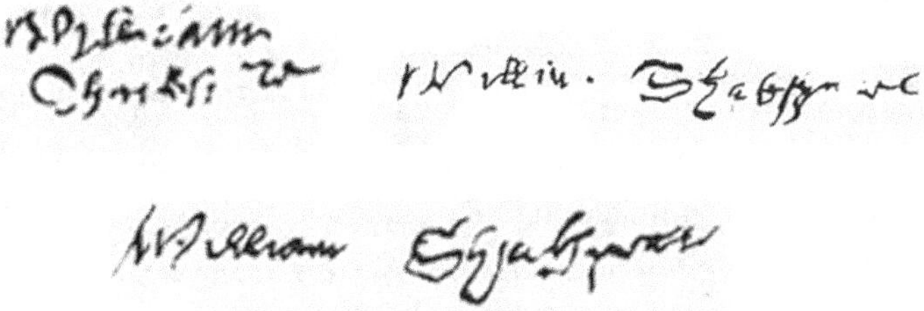

Signatures on the the Will, March 25, 1616

Figure 2. The six known signatures of William Shakspere of Stratford.

his will, which was dated March 25, 1616, but which may have been in preparation for some months.

These six signatures hardly give the impression of someone who lived by the pen, creating poems and plays for a total of more than 880,000 words. Jane Cox, who was Custodian of the Wills at the Public Records Office in London, wrote:

It is obvious at a glance that these signatures, with the exception of the last two [the Blackfriars signature, Nos. 2 and 3] are not the signatures of the same man. Almost every letter is formed in a different way in each. Literate men in the sixteenth and seventeenth centuries developed personalized signatures much as people do today . . . (Michell, 1996, p. 100)

Anyone who has not been indoctrinated with the orthodox beliefs concerning Shakespeare may be rather puzzled by the fact that this man, who is credited with writing almost a million words, never developed a recognizable signature. Some independent scholars point out that, in the 16th and 17th centuries, it was normal practice for a law clerk to sign on behalf of a client who was illiterate. The client had merely to touch the signature and attest that that was indeed his name.

The proposed portraits of "Shakespeare" are a major puzzle. An early image of William Shakspere is based on a sketch of a monument to "Shakspeare," erected in Holy Trinity Church at an unknown date. This sketch was made by the antiquarian Sir William Dugdale in July 1634. An engraving, based on Dugdale's drawing, was prepared by Wenceslaus Hollar and included in Dugdale's *Antiquities of Warwickshire* published in London in 1656.

The earliest purported image we have of the great author is that prepared by Martin Droeshout for inclusion in the publication, in 1623, of *Mr William Shakespeare's Histories Comedies and Tragedies*, now referred to as the First Folio. This image, which is shown in Figure 3, obviously bears little or no relationship to the Dugdale portrait shown in Figure 4. There is no record of what model—if any—Droeshout used in preparing his engraving.

A number of scholars have listed a number of problems with the Droeshout portrait. See, for instance, David Roper (2008, p. 408 et seq.). For instance, the head of the figure is too large for the body. Another cause for concern is the *thick line that extends from beneath the chin, upwards to the lobe of the left ear, which looks suspiciously like the outline of a mask.*

The image of Shakespeare that one can see today (Chiljan, 2011) is shown in Figure 5. This image appears to be that of a well-fed and self-satisfied man whose hands rest on a cushion, the right hand holding a quill and the left hand resting on a small sheet of paper. This version of "Shakespeare" obviously bears little or no resemblance to either the portrait sketched by Sir William Dugdale in 1634, or to the Droeshout portrait featured in the First Folio. Bianchi (2018) claims to find evidence that this bust was installed in Holy Trinity Church in 1746, in the course of repairs (replacing an older bust by Gerard Johnson), and that the new bust is actually that of Carlo Vizziani (died 1661), an Italian attorney who was Rector of La Sapienza, of the University of Rome.

Figure 3. Martin Droeshout's portrait engraving of Shakespeare on the title page of the First Folio (1623).

Figure 4. Hollar's engraving of Sir William Dugdale's portrayal of the Shakespeare monument (July 1634).

Figure 5. Shakespeare monument as it appears today in The Holy Trinity Church, by Gerard Johnson (but see Bianchi, 2018).

Figure 6. The "John Sanders" portrait, ostensibly of William Shaksper (1603).

Quite recently, what is now known as the "Sanders portrait," shown in Figure 6 (Sanders portrait, Wikipedia, 2020; Nolen, 2010), has been proposed as a portrait of "William Shakespere." This portrait is currently owned by Lloyd Sullivan, who is believed to be a distant relative of John Sanders, who is believed to have been an early (perhaps the first) owner of the portrait, and who may have been the painter. The painting has been in the same family for 400 years. However, the clothing includes silver threads, which were worn only by noblemen at that time. A rag-paper label, that was attached to the back of the portrait at an unknown date, carries text that is now illegible but was transcribed in 1909 as follows:

> *Shakespere*
> *Born April 23–1564*
> *Died April 23–1616*
> *Aged 52*
> *This likeness taken 1603*
> *Age at that time 39 y$^{s.}$*

The quoted date of birth is consistent with the recorded date of baptism (April 26, 1564), and the quoted date of death is consistent with the recorded date of burial (April 25, 1616). The portrait, painted in oil on an oak panel, has been subjected to many tests none of which—to date—invalidates the proposed credentials of the portrait.

To sum up, there is no accepted portrait of William Shakspere, and the images we do have are grossly inconsistent. The Sanders portrait has the merit that it could be a real portrait of a real person that was prepared while the subject was alive in the 16th Century.

In scientific research, it is always good to have more than one hypothesis in mind. The legal system would not work very well if the judge were required to listen to the attorney for the prosecution, and to ignore the attorney for the defense. So if we are to consider—or reconsider—

the case for William Shakspere as the great author, we should pay some attention to one or more alternative candidates. If all alternatives fare worse than Shakspere, then the case for the orthodox candidate will be strengthened. If, on the other hand, one of the alternatives is found to have a stronger case to present, that would be a good reason to reconsider one's support of William Shakspere. For this reason, we now turn our attention to the current leading alternative candidate for the title of Author. He is Edward de Vere, 17th Earl of Oxford.

3. EDWARD DE VERE, 17TH EARL OF OXFORD, THE LEADING CHALLENGER

For at least three hundred years, various scholars have—for various reasons—sought an alternative identity for the great author we know as *Shakespeare*. This search obviously reflects a dissatisfaction with the orthodox candidate, William Shakspere. Some of the reasons for this dissatisfaction were evident in the preceding section. We shall find further reasons in subsequent sections. A few of the alternative candidates were listed in the **Introduction**, where we named Edward de Vere, 17th Earl of Oxford, as the current favorite.

Since de Vere was a nobleman, there is of course a great deal of information about him in the public record. Yet—oddly enough— there is also a good deal of information that is conspicuously missing. For instance, the circumstances of his death are uncertain. There was an uncanny silence about it. There was no grand funeral. There was no public mourning. No one wrote a eulogy concerning a premier nobleman (who may have been the most famous poet and playwright of the time—or perhaps of all time).

Edward de Vere was born on April 2, 1550, at the de Vere ancestral home, Hedingham Castle, in Essex. During his father's lifetime, Edward had the courtesy title (not an official title) of *Viscount Bolebec*. He began his remarkable education very early, first with tutors, then becoming a student at Queen's College, Cambridge, at the tender age of eight. The principal known facts about Oxford's life are listed in Table 1, which is shown at the end of the article.

The 16th Earl died in 1562, whereupon de Vere became the 17th Earl, inheriting the earldom's estates and the title of Great Lord

Chamberlain, becoming the premier earl in the country and the richest. Since Oxford was underage (12), he became a royal ward and was assigned to the care of Sir William Cecil (later Lord Burleigh) whose estate was on the Strand. It is significant that the Cecil home had one of the most extensive libraries in Europe.

Oxford had excellent tutors (Thomas Fowler, Lawrence Nowell, and Sir Thomas Smith), and would have had an association with the great scholar Arthur Golding (his uncle), who was in the employ of Burleigh. Golding is known as a translator of Ovid's *Metamorphoses* which had a great influence on young Oxford. (Some suspect that Oxford actually prepared the translation.) Oxford became fluent in Latin and French, and probably competent in one or two other European languages such as Spanish and Italian. After a brief widowhood, de Vere's mother was remarried (to Sir Charles Tyrell), an event that has a strong echo in *Hamlet*, which some scholars suspect to be autobiographical.

At the age of fourteen, Oxford was registered as a student at St John's College, Cambridge. At the age of seventeen, Oxford was admitted to Gray's Inn for legal studies which, some scholars suspect, led to the extensive and remarkably accurate knowledge of the law in the works of Shakespeare. At about that time, Oxford by accident killed Thomas Brincknell, a servant in the home of William Cecil, while practicing fencing maneuvers with another employee, Edward Baynom. Brincknell was considered to have been drunk at the time, and the jury returned a verdict of *felo de se* (death by his own fault).

Oxford was a skilled dancer and very witty. Not everyone at court appreciated his wit, but he had overriding protection since he became a favorite of the Queen, who called him her "Turk." Oxford was keen to engage in military service, which was the normal ambition of a nobleman. The Queen routinely refused his petitions, but he did serve briefly under the Earl of Sussex in putting down the rebellion of the Northern English Catholic nobles, and was part of the fleet that sailed out to confront (and defeat) the Spanish Armada.

Oxford was highly athletic and distinguished himself in several tournaments, which further raised his status in the eyes of the Queen.

In 1571, Oxford married Anne Cecil, the fifteen-year-old daughter of William Cecil. To make this marriage possible, the Queen raised Cecil to the peerage with the title Lord Burleigh. Most marriages among the nobility were not love-matches, and this marriage proved to be rocky on the part of Oxford, although Anne was loyal and loving throughout.

In 1574, still anxious to distinguish himself with military service, Oxford went (without leave) to Flanders with the goal of taking part in the military campaign against Spain, but the Queen soon had him escorted back to England.

In 1575, the Queen finally gave Oxford leave to travel, which he did *con brio*. He traveled to Paris, where he was received with honor at court, then went on to Strasbourg, where he met the great scholar Sturmius. Then began his year-long travels in Italy, with which Oxford became enthralled. He set up home for some months in Venice, but also visited Florence, Genoa, Mantua, Milan, Padua, Siena, Verona, and possibly Messina and Palermo in Sicily. He adopted Italian manners and dress—so much so that on his return to England he was referred to as *the Italianate Englishman*.

The word *economy* never entered into Oxford's lexicon, and he instructed Burleigh to sell his estates whenever necessary to cover his expenses. So began his descent into penury.

While he was in Italy, his wife gave birth to a daughter, Elizabeth. However, Oxford learned of rumors that he was not the father of this child. Hot-headedly, he refused to meet with her or her relatives who were waiting to greet him on his return to England. Oxford distanced himself from Anne who continued to live with their daughter at the home of Burleigh (Anne's father).

In 1580 Oxford purchased a mansion known as *Fisher's Folly* in Bishopsgate, where he is reputed to have set up a "college" for aspiring poets and playwrights, including Thomas Churchyard, Thomas Lodge, John Lyly, Anthony Mundy, Thomas Nashe, and Thomas Watson.

In 1581, Oxford confessed to the Queen his involvement with a Catholic party and was sent briefly to the Tower of London. He was later reconciled with his wife who subsequently bore him a son (born and died in 1583) and three daughters, one of whom died in infancy.

Not surprisingly, Oxford had a mistress (reputedly a dark-complexioned beauty) named Anne Vavasour. Anne had a miscarriage in 1580, and gave birth to a son on March 21, 1581. However, Anne was a lady-in-waiting to the Queen, who was not amused and sent Anne, Oxford, and their son, to the Tower. They were released on June 8.

Oxford remained out of favor with the Queen until June 1, 1583, when he was finally allowed to return to court. This would have been a period of disgrace, such as one may find as a feature of the Sonnets. Anne's uncle, Thomas Knivet, took umbrage at Oxford's dishonoring his family, and there began a feud between the two families, sometimes acted out in a manner reminiscent of the feud between the Montagues and the Capulets. In one of the encounters, Knivet succeeded in wounding Oxford (which may be related to references to the *lameness* of the author in the Sonnets).

Oxford's financial situation went from bad to worse. For instance, he had invested (and lost) £3,000 in Frobisher's third attempt to find a Northwest Passage. To everyone's surprise, the Queen (usually very tight-fisted) granted Oxford a lifetime annuity of £1,000, payable quarterly, with no accounting required. This atypical act of generosity remains unexplained.

Oxford's wife Anne died in 1588. With the Queen's blessing—and perhaps at her instigation—he married one of the Queen's maids of honor in 1591. Fortunately, Oxford's new wife, Elizabeth Trentham, was wealthy. She bore him a son and heir, Henry, in 1593. (Surprisingly, Oxford seemed to show little interest in his son.) Oxford's daughter Elizabeth married the sixth Earl of Derby in 1594. Oxford's daughter Susan married Philip Herbert, later 4th Earl of Pembroke and 1st Earl of Montgomery, one of the "incomparable pair of brethren" to whom the First Folio was later dedicated. The "incomparable pair" were brothers Philip and William Herbert, sons of Countess Mary Sidney Herbert, thought to be the most educated woman in England at the time, comparable to the Queen.

From 1591 on, apart from his participation in state trials, etc., little is known of Oxford's life except that he patronized literature and supported a company of actors. He was acclaimed by his contemporaries as the "best playwright" of the time but writing under an alias (Table 1).

In 1596, Oxford's wife purchased a house known as King's Place in Hackney, then a village near to the capital. It is believed that Oxford died at Hackney on June 24, 1604, and was buried at St. Augustine's Church. An entry in the church register has the annotation "plague." However, scholars find it curious that there was no memorial, and (as far as we can tell) Oxford left no will.

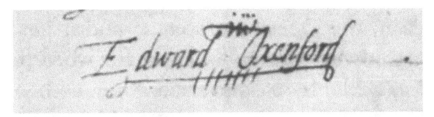

Figure 7. A typical signature of Lord Oxford.

For comparison with the purported signatures of William Shakspere shown in Figure 2, we show in Figure 7 a typical signature of de Vere. The symbol just above the gap between *Edward* and *Oxenford* is thought to indicate a coronet, indicative of his status as Earl. A sample of de Vere's penmanship is shown in Figure 8. This is a letter written (in French) when he was in his teens.

Monsieur treshonorable 79 N°25.

Monsieur i'ay receu voz lettres, plaines d'humanité et courtoysie, et fort resemblantes a vostre grand amour et singuliere affection envers moy. comme vrais enfans devëment procreez d'une telle mere. pour la quelle ie me trouve de iour en iour plus tenu a V. h. voz bons admonestements pour l'observation du bon ordre selon voz appointemens, ie me delibere (dieu aidant) de garder en toute diligence, comme chose que ie cognois et considere tendre especialement a mon propre bien et profit, usant en cela l'advis et authorité de ceux qui sont aupres de moy. la discretion desquels i'estime si grande (s'il me convient parler quelque chose a leur advantage) qui non seulement ilz se porteront selon qu'un tel temps le requiert, ains que plus est feront tant que ie me gouverne selon que vous aves ordonné et commandé. Quant a l'ordre de mon estude pour ce que il requiert un long discours a l'expliquer par le menu, et le temps est court a ceste heure, ie vous prie affectueusement m'en excuser pour le present. vous asseurant que par le premier passant ie le vous feray scavoir bien au long. Cependant ie prie a dieu vous donner santé.

Edward Oxinford

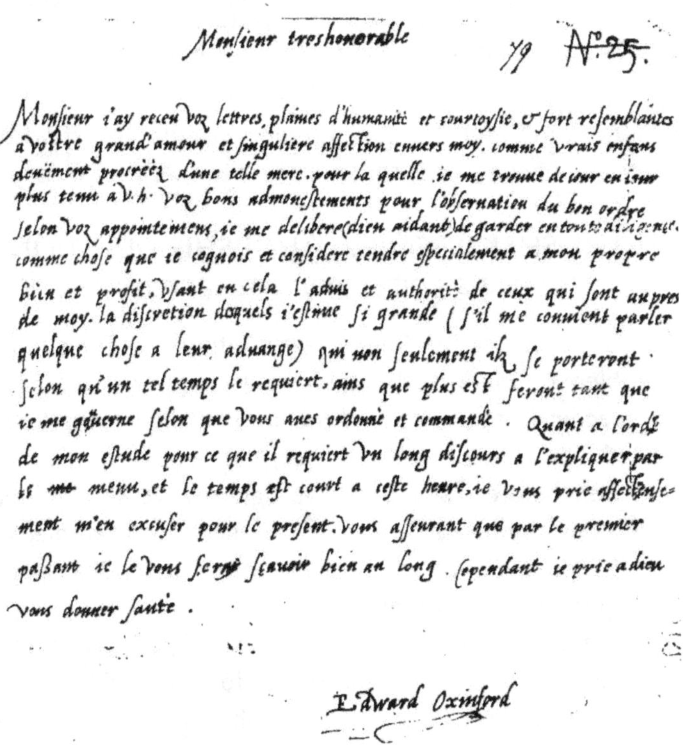

Figure 8. A sample of Lord de Vere's penmanship in a signed letter written in French when he was in his teens.

Figure 9. Edward de Vere, circa 1575. The Welbeck portrait, painted while Oxford was in Paris. National Portrait Gallery, London.

Figure 10. Portrait of Edward de Vere by Marcus Gheeraedts, known as the St. Albans portrait. Date unknown.

We show, in Figures 9 and 10, two portraits of Oxford that are believed to have been painted when he was twenty-five years old.

4. THE SHAKESPEARE AUTHORSHIP QUESTION FROM A SCIENTIFIC PERSPECTIVE

Our goal is to address the Shakespeare Authorship Question as if it were a problem of science rather than literature (Sturrock, 2013), and in that way specifically to understand the implications for the Authorship Question of discovering one or more cryptograms.

However, before discussing cryptograms, we should note that there are other significant forms of evidence. For example, we may consider the question of whether or not William Shakspere from Stratford-upon-Avon was a writer.

The independent scholar Diana Price (see Figure 11) has compiled evidence relevant to this question in the form of a *Chart of Literary Paper Trails* (Price, 2012). Price compares what is known about William Shakspere with what is known about twenty-four writers who lived in England at the same time as Shakspere. For each of these writers, and for Shakspere, we may follow Price in considering whether or not there exists evidence in each of ten categories relevant to the literary profession (Sturrock, 2008). We find that there is evidence conforming to at least three categories for each comparison author, but none for Shakspere. Our analysis of this evidence leads to the conclusion that there is only one chance in 100,000 that Shakspere was a writer (obviously implying that the Great Author was someone other than Shakspere).

Figure 11. Diana Price

In order to pursue the Authorship Question according to the guidelines of scientific inference, we adopt the terminology and methodology of an article entitled "Applied Scientfic Inference" (Sturrock, 1994), which is based on Bayesian principles. We may start by identifying a set of hypotheses that is *complete* in the sense that one and only one of the hypotheses must be true. We may then update our assessments of those hypotheses in response to the available relevant information.

One possible set of hypotheses would be

H1: *Shakespeare was Shakspere,* and
H2: *Shakespeare was not Shakspere,*

where *Shakespeare* denotes the *Great Author*.

We would need to update our assessments of these hypotheses in response to the results of the cryptogram analyses that we shall carry out in later sections. This would require us to decide how likely we are to find a cryptogram on the basis of each of these hypotheses. In order to relate our analysis to cryptograms, it is more helpful to adopt the following hypotheses:

H1: The Authorship Issue involved secrecy, and
H2: The Authorship Issue did not involve secrecy.

To find a cryptogram would obviously support H1. The whole purpose of a cryptogram is to send a message secretly. If there is no secrecy, there is no point in creating a cryptogram.

However, *according to the orthodox Stratfordian theory,* there was nothing secret about the identity of the author. There was no Conspiracy of Silence to hide the identity of the Great Author. Hence finding a cryptogram would support hypothesis H1. But since H1 is incompatible with the orthodox Stratfordian theory, finding a cryptogram comprises evidence against the orthodox Stratfordian theory.

When we come to analyze cryptograms, we shall be able to calculate the probability that the relevant text occurred by chance. Disproving chance (or showing that chance was unlikely) leads to the probability that the text had been created intentionally, which would rule out the orthodox Stratfordian theory. Hence the probability that a cryptogram has not occurred by chance can be interpreted as the probability that secrecy was involved, which may in turn be interpreted as the probability that the Stratfordian theory is false. Hence if we choose to limit our choices to the two hypotheses

> *Shakespeare was William Shakspere,* and
> *Shakespeare was Edward de Vere,*

then finding a cryptogram will represent evidence in support of the Oxfordian hypothesis.

If we were considering a standard laboratory experiment, for which the possible outcomes are expected to be well-known and for which the relevant theory is well-established, we could set probabilities (known as the "prior probabilities") on the possible results of the experiment before the experiment is undertaken. If the actual results are found to conform to the expectation, that would of course support the theory—otherwise not. However, the study of cryptograms is not the same as a standard laboratory experiment: *One does not know all the possible outcomes in advance.* This means that *one cannot treat the study of cryptograms in exactly the same way that one would treat the analysis of a laboratory experiment.*

In order to clarify the difference, it is helpful to revise our terminology. In the study of a laboratory experiment, one may expect to have enough information to assess the probability of finding each of the possible outcomes of the experiment. These are expressed as the *prior probabilities.* However, anyone looking for hidden messages has at best only a vague idea of what they might find, and may therefore have only a vague interpretation of whatever text he might find more-or-less hidden in the material under investigation. One must expect that different analysts are likely to have different interpretations of whatever hidden messages they might find—or think they have found. To recognize this intrinsic—necessarily subjective—characteristic of cryptology, it seems helpul to introduce the term *degree of belief*, for which we use the notation *DOB* (Sturrock, 2013).

This concept (*degree of belief*) plays the same role in the analysis of cryptograms, etc., as the concept of probability does in the analysis of laboratory experiments. So we would start with a prior *degree of belief* that the Authorship Issue involved secrecy, and a prior degree of belief that it did not, etc. Then we need to adjust that *degree of belief* in reponse to whatever evidence we find concerning cryptograms, etc.

As we shall see, some of these *degrees of belief* can be very small. In the usual notation for a probability, one might be encountering and combining numbers like 0.001 and $5 \; 10^{-6}$. Besides being a little awkard to deal with, it is not too easy to "get a feel" for such numbers.

In an article on *Applied Scientfic Inference* in this journal (Sturrock, 1994), we have adopted a suggestion of Edwin Jaynes (2003), who pointed out that a concept that originated in electrical engineering can be very useful in the present context. We can measure a probability (or a *degree of belief*) in *decibels*, which has the abbrevation *db*. If we start by assigning a probability P to a proposition, this may be converted to an *Odds* by

$$Odds = \frac{P}{1-P}$$

(1)

The analyst can then express his *degree of belief* in the proposition as follows:

$$DOB = 10 \times \log_{10}\left(Odds\right)$$

(2)

The following Table 2 gives a few examples of this conversion.

Since this notation may not appeal to every reader, we shall normally express a *degree of belief* both as a probability and as measured in *db*.

TABLE 2
Relating Probability, Odds, and Degree of Belief

Probability	Odds	Degree of Belief in *db*
0.001	0.001	−30
0.01	0.01	−20
0.1	0.11	−9.5
0.5	1	0
0.9	9	9.5
0.99	99	20
0.999	999	30

5. AN INTRODUCTION TO THE DEDICATION
OF THE SONNETS AND ITS MESSAGES

Shake-Speare's Sonnettes was registered for publication by the Stationers' Company on May 20, 1609. The entry in the Stationers' Register reads

> Entred for his copie under thandes of master
> Wilson and master Lowndes Warden a Booke
> called SHAKESPEARES sonnettes.

The publisher was Thomas Thorpe, and the book was to be sold by two booksellers: William Aspley at the sign of The Parrot in St Paul's churchyard, and William Wright at Christ's Church Gate near Newgate. As Jonathan Bond (2009) has commented: *Of the birth in print of what would come to be the most celebrated poems in the English language, not another word was said.* The SONNETS disappeared.

The title page is shown as Figure 12 and the Dedication as Figure 13. The space between parallel lines on the title page would normally have contained the name of the author. For *Shake-Speare's Sonnettes*, the location is blank.

The Dedication receives little attention from orthodox Shakespeare scholars, perhaps because a dedication would normally be composed by the publisher. Most scholars infer from the initials

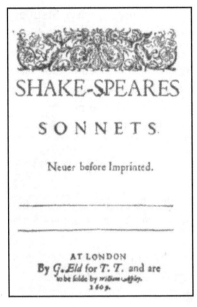

SHAKE-SPEARES

S O N N E T S.

Neuer before Imprinted.

AT LONDON
By G. Eld for T. T. and are
to be solde by william aspley.
1609.

TO. THE.ONLIE.BEGETTER.OF.
THESE. INSVING. SONNETS.
Mʳ. W. H. ALL.HAPPINESSE.
AND.THAT.ETERNITIE.
PROMISED.

BY.

OVR.EVER LIVING.POET.

WISHETH.

THE. WELL-WISHING.
ADVENTVRER . IN.
SETTING.
FORTH.

T. T.

Figure 12.   The title page of
Shake-speare's Sonnets.

Figure 13.   The Dedication in
Shake-speare's Sonnets.

"T.T.", in the bottom right corner, that the Dedication was composed by Thomas Thorpe. However, this dedication is unlike any other dedication of that era, and unlike any dedication composed by Thomas Thorpe, as we see from an example of a Thorpe dedication shown as Figure 14.

TO HIS KIND,AND TRVE FRIEND:
EDWARD BLVNT.

Lount: *I purpose to be blūt with you, & out of my dulnesse to encounter you with a* Dedication *in the memory of that pure Elementall wit* Chr. Marlow; *whose ghoast or* Genius *is to be seene walke the* Churchyard *in (at the least) three or foure sheets. Me thinks you should presently looke wilde now, and growe humorously frantique*

Thine in all rites of perfect friendship,

THOM. THORPE.

Figure 14.   Thomas Thorpe's (typical)
dedication of a book, to his
colleague Thomas Blount.

In a book of 490 pages entitled *Shake-speare's Sonnets*, editor Katherine Duncan-Jones (1997) reproduces the Dedication, remarks that *the over-rhetorical wording is evidently Thorpe's*, and comments on what she assumed was Thorpe's description of himself as *THE WELL-WISHING ADVENTURER*.

In a book of 493 pages, also entitled *Shakespeare's Sonnets*, Stephen Booth (2000) refers to *Thorpe's dedication* (p. 547) but neither reproduces nor discusses it.

In a book of 671 pages entitled *The Art of Shakespeare's Sonnets*, Helen Vendler (1997) does not even mention the Dedication.

These orthodox scholars, who naturally believed the poems to be the work of William Shakspere, never suspected that the Dedication might contain one or more hidden messages.

What attention the Dedication has received from orthodox scholars has been speculation about the identity of "Mr. W.H." According to Stanley Wells (1970, p. 6), *"Mr W.H." provides the biggest puzzle of all.* According to Samuel Schoenbaum (1970), *the identity of "Mr. W.H." is a riddle that to this day remains unsolved.*

The fact that the Dedication actually contains hidden messages was discovered not by a Shakespeare scholar but by a physicist and electrical engineer—John Rollett (see Figure 15).

John M. Rollett studied physics at Trinity College, Cambridge, and received a PhD degree from London University. He was for many years an engineer at the British Post Office Research Station at Dollis Hill in northwest London, and was author of about fifty articles and patents. Dollis Hill was the principal research station in Britain for telephones and related technology. Rollett was closely involved in the major Post Office project at that time—the design and installation of a new transatlantic telephone cable. He was known to his colleagues as highly intelligent and highly inquisitive, and was known for his persistence in sticking with a difficult problem until it was solved. Rollett had wide interests, including Elgar's *Enigma Variations* on which he wrote a short book, and he would discuss these interests at length with his Dollis Hill colleagues.

As we shall see, Rollett was responsible for a breakthrough in Shakespeare Authorship research, which contributed to the current pre-eminence of the candidacy of Edward de Vere, Earl of Oxford. However, ever an independent scholar, Rollett later advocated

Figure 15. John M. Rollett

William Stanley, Earl of Derby, as the great writer we know as Shakespeare (Rollett, 2015).

There is no better introduction to the mystery of the *Dedication of the Shakespeare Sonnets* than the one written by Rollett himself (Rollett, 1999, 2004, pp. 253–266), which followed his seminal articles in 1997 (Rollett, 1997a, 1997b).

> *There it is, so familiar, and so obscure: what an amazing production! There's nothing like it anywhere else in Elizabethan or Jacobean literature. What does it mean, for a start? What is it trying to tell us? The opening phrase is so well known, "To the onlie begetter," but how many people know that the spelling of "onlie" is very rare indeed? It could have been, in its tiny way, a clue to something quite unexpected until very recently. Surely there is rather more to the Dedication than first meets the eye.*

It is interesting to see how Rollett was led to his discovery. In 1964, "the 400th anniversary of a certain gentleman from Stratford," more than 400 books dealing with Shakespeare were published. The Shakespeare scholar Leslie Hotson published a book entitled "Mr. W.H.," in which Hotson claimed to have determined the identity of "Mr. W.H." (Hotson, 1964). Rollett initially found the book "completely convincing." Hotson's proposal was that "W.H." referred to William Hatcliffe, who was admitted as a law pupil to Gray's Inn in 1586. The next year, Hatcliffe was elected Prince of Purpoole, "a kind of temporary Lord of Misrule or Lord of Liberty," to preside over the festivities of the Christmas Season. In that position, Hatcliffe would have been expected to act like a prince of royal blood. Had the festivities included an induction ceremony in which Hatcliffe was carried on a throne covered by a canopy, it might have explained the opening lines of Sonnet No. 125, "Were it ought to me I bore the canopy . . . " However, Rollett learned (and Hotson should have known) that no canopy was ever carried over a Prince of Purpoole.

Hotson declared that the Dedication was a cryptogram composed by Thorpe. His interpretation involved a complex procedure—He starts with "Mr. W.H." in line 3, moves down to pick up the H in the next line, chooses HAT from this word, then drops down to line 7, and picks up

EVER-LIV-ING. In this way, Hotson picked up "HATLIV", *which seemed to be a reasonably good shot at "Hatcliffe."* Rollett initially in 1964 accepted this argument, but by 1967 he decided that it was "utter nonsense." For Rollett, *there were too many arbitrary steps in the proposed solution.*

Rollett remarked: *It [was] obvious that Hotson was very strongly biased towards the result he claimed to find.* He added, *It is not a good idea to have preconceptions in this kind of endeavor.*

However, the time that Rollett had spent in following Hotson's trail led him to suspect that, although Hotson's proposed cryptogram was nonsense, the Dedication seemed to be strange enough that it might be concealing some kind of message. Rollett noted that one of the oddest features of the Dedication is the full-stop after every word. It occurred to Rollett that this suggested that one should count words . . . for instance, every 3rd word, or every 5th word, etc. That idea led nowhere, so Rollett then tried alternating numbers—e.g., every 3rd word, then every 5th word, etc. That also led nowhere.

Rollett then focused on another peculiarity of the Dedication: The text is laid out in three inverted pyramids, of lengths 6 lines, 2 lines, and 4 lines. Perhaps the message (if there was one) could be found by taking the 6th word, then the 2nd word, then the 4th word, and repeating. This led Rollett to the sequence

THESE SONNETS ALL BY EVER

Actually, the complete cryptogram reads

THESE SONNETS ALL BY EVER THE **FORTH**

We examine this discovery in the next section and in Section 10.

6. "THESE SONNETS ALL BY EVER THE FORTH"

Rollett was intrigued with the discovery of

THESE SONNETS ALL BY EVER THE FORTH

in the Dedication. However, Rollett had never heard of an Elizabethan poet named *EVER*, leading him to dismiss the idea that the Dedication might contain a cryptogram.

Two or three years later, Rollett was in a library and on an impulse decided to look up the article on Shakespeare in the *Encyclopedia Britannica*. Towards the end of the article, he found a section headed "Questions of Authorship." He read the general arguments, including a paragraph on Francis Bacon, then came to the following two sentences:

> *A theory that the author of the plays was Edward de Vere, 17th earl of Oxford, receives some circumstantial support from the coincidence that Oxford's known poems apparently ceased just before Shakespeare's works began to appear. It is argued that Oxford assumed a pseudonym in order to protect his family from the social stigma then attached to the stage, and also because extravagance had brought him into disrepute at Court.*

Rollett immediately recalled the word EVER, and realized that it could be read as E VER for *Edward Vere*.

However, this discovery also made no great impression on Rollett. He was still looking for the identity of "Mr. W.H.", and still did not doubt that the gentleman from Stratford-upon-Avon was the author of the Sonnets and everything else. It was, as he remarked (Rollett, 1999), *A strange coincidence, not to say a thought-provoking one, but I still remained very skeptical, and was sure that chance was the most likely explanation of this odd result.*

Rollett noted that there was a possible connection between de Vere and the Dedication in that the sequence 6 – 2 – 4 matches the number of letters in the name Edward de Vere. Nevertheless, Rollett was disappointed that this sentence still did not seem to make sense. He could find no way in which *de Vere* was the "fourth" in anything.

The true meaning of "the forth" or "the fourth" may never be known, but Jonathan Bond, whom we shall meet in the next section, has offered the following suggestion:

> *de Vere, on reaching his majority, was keen to undertake military service, but the Queen for some time refused that request. Had she given approval, Oxford's military service would most likely have been in the Netherlands, where the Protestant population was waging war against the occupying power, Spain. England*

was not officially involved in that struggle until Antwerp was captured by Spanish forces in 1585. This led Elizabeth to sign the Treaty of Nonsuch, which brought England into the war against Spain in support of The Netherlands. De Vere was then allowed to engage in military service—but only briefly.

Bond has pointed out that that the Dutch for "the fourth" is "de vierde," which is phonetically close to "de Vere." This suggestion is intriguing. There may be no persuasive interpretation of "The Fourth" that we can identify four centuries after the Dedication was composed. It is possible that "the fourth" was part of an in-joke between the author of the Dedication and the intended recipient. de Vere may have been the fourth "something" that had some special significance for de Vere and the dedicatee. There is some indication that de Vere was the fourth-ranking member of the Queen's Privy Council, which may have given him some leverage in negotiations with the Queen and Robert Cecil.

We return to our discussion of the possible significance of "The Forth" in Section 10.

This discussion hinges on the question of whether or not the sentence THESE SONNETS ALL BY EVER THE FORTH was intentionally built into the Dedication, or appeared purely by chance. We can address this question by supposing that the author went through many versions, using the same words but in many different arrangements of those words. For instance, we can leave the words in their actual order, but change the rule for selecting words. Rather than select the 6th word, then the 8th word, then the 12th word, etc., we suppose that we can select any seven words. Then, keeping them in the order in which they actually occur, we can examine the sequence for a sensible message. None of them looks like a sensible message.

We have actually carried out one thousand simulations, and the four that seem nearest to a sensible message are the following:

OF THESE SONNETS W H HAPPINESSE PROMISED
THE INSUING Mr EVER WELL WISHING ADVENTURER
ONLIE W HAPPINESSE OUR POET WELL WISHING
ONLIE W PROMISED THE ADVENTURER SETTING FORTH

THE THESE INSUING H ETERNITIE BY POET
THE HAPPINESSE THAT ETERNITIE BY POET WISHING
ONLIE OF W HAPPINESSE AND THAT POET
ONLIE H AND THAT OUR WISHING FORTH
BEGETTER H HAPPINESSE BY EVER WISHETH THE
OF INSUING SONNETS W AND WELL FORTH
PROMISED BY OUR EVER POET WISHING FORTH
THE THESE ALL THAT BY POET FORTH
OF THESE ETERNITIE PROMISED EVER THE WELL
ONLIE OF THAT ETERNITIE OUR EVER POET
THE OF INSUING Mr H BY WISHETH
Mr HAPPINESSE AND PROMISED POET WISHING SETTING
ONLIE THESE HAPPINESSE LIVING ADVENTURER IN SETTING
THE INSUING SONNETS H ETERNITIE EVER WISHETH
BEGETTER OF ALL ETERNITIE BY EVER LIVING
TO THESE THAT ETERNITIE LIVING THE ADVENTURER
OF INSUING ETERNITIE WISHETH THE WISHING FORTH
BEGETTER AND PROMISED LIVING WISHETH THE FORTH
THE Mr W ETERNITIE THE WELL WISHING
TO H ALL EVER WISHETH ADVENTURER FORTH
TO OF Mr OUR THE WISHING FORTH
SONNETS W ALL THAT EVER ADVENTURER IN
ONLIE SONNETS AND THAT THE WELL WISHING
INSUING HAPPINESSE ETERNITIE LIVING WISHETH THE WISHING
INSUING Mr AND THAT WISHETH WISHING IN
HAPPINESSE AND PROMISED BY WISHETH THE ADVENTURER
THE H THAT ETERNITIE LIVING WELL SETTING
W ALL PROMISED BY LIVING WISHETH IN
TO ONLIE INSUING ETERNITIE WELL WISHING FORTH
TO INSUING Mr ETERNITIE THE WELL IN
OF Mr H POET THE IN SETTING
TO INSUING Mr ALL ETERNITIE BY EVER
THE THESE Mr HAPPINESSE THAT ETERNITIE WISHETH
TO SONNETS ETERNITIE PROMISED OUR THE FORTH
ONLIE BEGETTER THESE INSUING SONNETS ALL LIVING
TO BEGETTER SONNETS Mr H LIVING FORTH
TO ETERNITIE PROMISED OUR EVER POET ADVENTURER
INSUING THAT ETERNITIE LIVING WELL WISHING SETTING
TO ONLIE W ALL ETERNITIE OUR EVER
TO THAT BY OUR THE WELL SETTING
THE W AND ETERNITIE THE WISHING SETTING
TO THE ONLIE PROMISED WISHING IN SETTING
ONLIE BEGETTER SONNETS Mr ETERNITIE BY POET
BEGETTER H THAT OUR POET WELL WISHING
TO BEGETTER INSUING THAT ETERNITIE EVER FORTH
ONLIE OF INSUING HAPPINESSE WISHING IN SETTING
TO OF INSUING SONNETS HAPPINESSE WELL IN
TO Mr OUR LIVING POET WISHING ADVENTURER
TO THE BEGETTER THAT OUR WISHETH SETTING
BEGETTER INSUING Mr HAPPINESSE POET SETTING FORTH

Figure 16. Fifty of the random sentence simulations of the message "THESE SONNETS ALL BY EVER THE FORTH."

We give a list of 50 such "sentences", obtained by this random procedure, in Figure 16.

However, none of these is both grammatical and meaningful. The implication of this experiment seems to be that there is less (probably very much less) than one chance in a thousand (DOB less than −30) that the sentence THESE SONNETS ALL BY EVER THE FORTH occurred in the Dedication by chance. It would appear that THESE SONNETS ALL BY EVER THE FORTH was built into the Dedication by intent, the author first deciding on this sentence as one he would like to include (and conceal), then building text around it.

Who was the author of the Dedication? Since the letters T. T. appear in the bottom-right-hand corner of the page, scholars instinctively assume that the author was the publisher Thomas Thorpe, and the Dedication is generally referred to as "Thorpe's Dedication." However, once we realize that the Dedication contains a secret message, we obviously need to reconsider that assumption. Why should the *publisher* want to inform us that *These sonnets [are] all by EVER the Fourth (or the Forth)*?

If the Dedication was not composed by Thomas Thorpe, then who did compose it? The obvious answer is *Ever*—Edward de Vere. Although it may be normal practice for a dedication to be composed by the publisher, there is nothing normal about this Dedication—as we shall see in subsequent sections.

7. FINDING "HENRY WRIOTHESLEY"

The objective of the innocent letter code, Rollett found, is

> to distribute the words of the secret message systematically throughout the words of what seems a normal letter. . . . It was used . . . by prisoners of war in World War II, notably those in Colditz Castle sending information about the German war effort back to the UK.

There has to be a "key" to unlock the message, and various schemes have been devised. As an example, "Dear George" contains ten letters, and the key might be to select every tenth word. One way to read the hidden message would be to prepare a grid in which the first row comprises letters 1 to 10, the second row comprises letters 11 to 20, and so on. Then the hidden message would be found by reading the columns so formed. One (or more) of the columns would reveal the message. It might be necessary to read the column from top to bottom or vice versa.

This is an example of what is known as *Equidistant Letter Sequencing* (*ELS*) in which the text is rearranged into a rectangle or "grid", and the hidden message is revealed by reading the contents of the columns.

This chain of thought led Rollett to count the number of letters in the Dedication. He found that the Dedication contains 144 letters. This caught his attention, since it is possible to arrange 144 letters in a number of rectangles: 8 × 18, 9 × 16, 12 × 12, etc.

As he writes (Rollett, 1999, p. 68),

> *The first thing I noticed was in the array with 15 letters in each row, HENRY! [see Figure 17]. It is evident that the letters of the name are all equally spaced – every 15th letter starting from the H spells out the name. This is sufficiently unusual to suggest that it might have been deliberately arranged by the cryptographer. But Henry who? . . . . Perhaps his name was "Henry Oliver," the surname being indicated by the letters OLVR which follow on*

T	O	T	H	E	O	N	L	I	E	B	E	G	E	T
T	E	R	O	F	T	H	E	S	E	I	N	S	U	I
N	G	S	O	N	N	E	T	S	M	R	W	H	A	L
L	H	A	P	P	I	N	E	S	S	E	A	N	D	T
H	A	T	E	T	E	R	N	I	T	I	E	P	R	O
M	I	S	E	D	B	Y	O	U	R	E	V	E	R	L
I	V	I	N	G	P	O	E	T	W	I	S	H	W	T
H	T	H	E	W	E	L	L	W	I	S	H	I	N	G
A	D	V	E	N	T	V	R	E	R	I	N	S	E	T
T	I	N	G	F	O	R	T	H						

Figure 17. Choosing every 15th letter of the Dedication spells the word HENRY.

> *down from HENRY, and I did look in various books to see if such a person flourished at the time, without success.*

Rollett continued to examine the various grids and, as he wrote (Rollett, 1999, p. 69),

Eventually the penny dropped. In the array with 18 letters in each row, I had repeatedly overlooked something. There, split up into three bits, is the name WR-IOTH-ESLEY, spelt perfectly, just as it was always spelt officially [see Figure 18]. I first noticed the letters ESLEY in the middle column, and almost immediately the letters IOTH in the one next to it. At that moment I knew with absolute certainty that I would find the letters WR somewhere, and there they are, at the bottom of the second column. Moreover, this is a perfect rectangle, where the cryptographer would naturally try to hide the most important information, since perfect rectangles are where a cryptanalyst would look first of all for something hidden. And if "onlie" had been spelt with an e between the n and the l,

T	O	T	H	E	O	N	L	I	E	B	E	G	E	T	T	E	R
O	F	T	H	E	S	E	I	N	S	U	I	N	G	S	O	N	N
E	T	S	M	R	W	H	A	L	L	H	A	P	P	I	N	E	S
S	E	A	N	D	T	H	A	T	E	T	E	R	N	I	T	I	E
P	R	O	M	I	S	E	D	B	Y	O	U	R	E	V	E	R	L
I	V	I	N	G	P	O	E	T	W	I	S	H	E	T	H	T	H
E	W	E	L	L	W	I	S	H	I	N	G	A	D	V	E	N	T
U	R	E	R	I	N	S	E	T	T	I	N	G	F	O	R	T	H

Figure 18. The 18 x 8 grid of the letters in the Dedication to the Sonnets shows the name WRIOTHESLEY in three pieces: WR in column 2, IOTH in column 11, and ESLEY in column 10.

as was usual, the number of letters would have been 145, with the wrong factors, so that particular e had to be omitted.

Rollett went on to calculate the probability that the name HENRY had occurred by chance, and that the combination WR-IOTH-ESLEY had also occurred by chance. We carry out these calculations in Appendices A and B. Our methodology is a little different from the one adopted by Rollett, but we arrive at a similar result. The probability that the name HENRY might have occurred by chance in one of the rectangular arrays is found to be 0.002, and the probability that the name WRIOTHESLEY might have occurred by chance, broken up in either two or three pieces, in one of the rectangular arrays, is found to be 7×10^{-6}. Hence the probability that the name HENRY WRIOTHESLEY might turn up by chance is approximately 10^{-8}—one chance in 100 million (DOB = –80).

Why Henry Wriothesley? That is the family name of the only *real* person mentioned by name in all of Shakespeare's plays and poems. His two somewhat erotic poems, *Venus and Adonis*, and *Rape of Lucrece*, are both dedicated to the Third Earl of Southampton, whose name was Henry Wriothesley. It is also significant that Wriothesley is widely believed to be the "Fair Youth" of the Sonnets. This discovery suggests that the enigmatic initials *WH* were originally *HW* for Henry Wriothesley. Whoever provided the Sonnets and Dedication to Thorpe may have considered it discreet to reverse the initials.

In pursuing our investigation of Shakespeare's Sonnets, we now meet another investigator—Jonathan Bond (Figure 19). There were two themes in Bond's life that gave him the skills and interests that led to his seminal contributions to Shakespearian cryptology—mathematics and the theater.

Figure 19. Jonathan Bond

Bond was born (in 1966) and raised in the North East of England—the same part of the country that produced the Shakespearian and Oxfordian scholar John Thomas Looney and the creative, and controversial, scientist (and author PS's one-time mentor) Fred Hoyle. In 1985, Bond became a student in the philosophy department of University College London, specializing in mathematical logic. Anyone who is willing to grapple with the intricacies of Gödel's theorem on incompleteness and undecidability in mathematics has more than adequate intellectual fortitude for investigations in cryptology.

Bond continued his mathematical studies at Cambridge University in 1991, where he also acquired an interest in acting, once playing the lead role in a Marlowe Society production of Peer Gynt. This experience sparked his decision to train as a professional actor at the Guildhall School of Music and Drama, which led to a twenty-year career in the theater. In 1997, Bond joined the Shakespeare Globe Theatre, where he played parts in a Beaumont and Fletcher play (*The Maid's Tragedy*) and (as Oliver) in *As You Like It*. This was the first of three seasons at the Globe, during which he appeared in *As You Like It*, *Midsummer Night's Dream*, and *Timon of Athens*. He has also appeared in *Romeo and Juliet* and *Twelfth Night* on the British stage. During his two years at the Globe, Bond worked alongside Mark Rylance. It was Rylance's inquiring spirit and fascination with the Authorship Question that sparked Bond's interest in the issues surrounding the composition of the "Shakespeare" plays.

With his mathematical background, Bond was drawn to the investigation of the Dedication of the Sonnets. There had been a number of suggestions concerning the Dedication in the literature, but no one had considered these suggestions all together. Bond thought that would be a useful undertaking.

One of the first things that Bond noticed is that the phrase "our ever-living" is very close to an anagram of the de Vere family motto *vero nil verius* (nothing truer than truth). It becomes an exact anagram if one replaces the final letter "g" with an "s". As Bond point out in *The de Vere Code* (Bond, 2009, p. 57), the capital letters S and G are very similar when written in "secretary hand," which was a standard script used in the 16th Century.

It is often the case in creative activity that it helps to step back for a while. This is when new insights may bubble up from the unconscious. So it was with Bond when, reviewing the literature once more in 2008 and 2009, he uncovered new evidence of encryptions in the Dedication, which he describes in *The de Vere Code* (2009): The 18 by 8 perfect rectangle grid has much more to reveal than Rollett had supposed.

What Bond noticed, as shown in Figure 20, is the sentence: TO ESPIE OFT WR-IOTH-ESLEY WIT NEED NOT HERE TRIE, or, in modern English, *To see Wriothesley often in these sonnets is easy if you use your wits.* This nine-word sentence appears in a perfect rectangle, reads left to right, is grammatical—and grammatically complex—and is spelled correctly. The forms of "espie," "oft," and "trie" are the appropriate spellings for an early 17th-century text.

As Bond writes,

> The ramifications of the full WRIOTHESLEY cipher are signifi-
> cant. The extent of the message takes it beyond conjecture into the
> realm of being . . . documentary proof that, unless the author of
> the Dedication was lying, Wriothesley was the primary subject of
> the sonnets. This in itself is a remarkable discovery, as important

T	O	T	H	E	O	N	L	I	E	B	E	G	E	T	T	E	R
O	F	T	H	E	S	E	I	N	S	U	I	N	G	S	O	N	N
E	T	S	M	R	W	H	A	L	L	H	A	P	P	I	N	E	S
S	E	A	N	D	T	H	A	T	E	T	E	R	N	I	T	I	E
P	R	O	M	I	S	E	D	B	Y	O	U	R	E	V	E	R	L
I	V	I	N	G	P	O	E	T	W	I	S	H	E	T	H	T	H
E	W	E	L	L	W	I	S	H	I	N	G	A	D	V	E	N	T
U	R	E	R	I	N	S	E	T	T	I	N	G	F	O	R	T	H

Figure 20. The 18 by 8 (perfect) grid reveals the complete sentence TO ESPIE OFT WR-IOTH-ESLEY WIT NEED NOT HERE TRIE.

to Stratfordians as to their opponents. . . . The WRIOTHESLEY cypher is so clearly not a coincidence as to be tantamount to proof that the author is encrypting messages in the Dedication. . . . The author . . . is playing a fantastically clever word game. And witty, too. . . . The encypherer loves word puzzles, and expects his reader to like them too. . . . If any doubts do remain as to his extraordinary skill in providing the recipient of the SONNETS with ever-more ingenious riddles to unravel, what follows puts the matter conclusively beyond doubt. Because, like any show-man worth his salt, the author saved his best tricks for last.

8. "PRO PARE VOTIS EMERITER"

John Rollett had examined the 12 by 12 grid and concluded that there was no message hidden there. Jonathan Bond for some time accepted Rollett's conclusion as a fact, but not without some hesitation.

The Dedication was rich in hidden messages, and the Author seemed to take great pleasure in giving the Recipient one treat after another. A recipient who suspects—or hopes—there is something hidden in a cryptogram from a friend or lover would be inclined first of all to examine the central grid. Rollett and Bond had, between them, discovered so much hidden in the Dedication, that it seemed to Bond very odd that there should be nothing hidden in the central grid.

Bond eventually decided to check it out. As Bond remarks,

I had taken Rollett's opinion at face value that there was nothing resembling a message in the most obvious perfect rectangle of all— the 12 by 12 square. I had always felt some unease about this. . . . Why *did he leave this one out? The answer—dare I say " unsurprisingly"— was that he didn't. On closer inspection, the reason why the message had not immediately been apparent became clear. **It is in Latin.***

The 12 by 12 grid, shown in Figure 21, contains the phrase *PRO PARE VOTIS EMERITER* all conjoined with that pivotal word *EVER*. What does it mean? There is a lot of flexibility in interpreting these words. Bond gives these guidelines:

PRO: Means *for* or *on behalf of* and their usual connotations in English.

PARE: Relates to the verb "pareo" with its primary sense of *"to come forth, appear, be visible, show oneself, to be present."* A second related form is "parens" meaning *parent*, or *procreator*. There is also a third form, used by Ovid [where] the word "pare" specifically means *companion, mate or consort.*

VOTIS: Means *to vow, promise solemnly, engage religiously, pledge, devote, dedicate, or consecrate.*

EMERITER: Relates to "emeritus" meaning *to obtain by service, gain, earn, merit, or deserve.*

T	O	T	H	E	O	N	L	I	E	B	E
G	E	T	T	E	R	O	F	T	H	E	S
E	I	N	S	V	I	N	G	S	O	N	N
E	T	**S**	M	**R**	W	H	A	L	L	H	A
P	**P**	**I**	N	**E**	S	S	E	A	N	D	T
H	**A**	**T**	**E**	**T**	E	R	N	I	T	I	E
P	**R**	**O**	**M**	**I**	S	E	D	B	Y	O	V
R	**E**	**V**	**E**	**R**	L	I	V	I	N	G	P
O	E	T	W	I	S	H	E	T	H	T	H
E	W	E	L	L	W	I	S	H	I	N	G
A	D	V	E	N	T	V	R	E	R	I	N
S	E	T	T	I	N	G	F	O	R	T	H

Figure 21. The Dedication set out as a 12 x 12 grid.

These rather flexible definitions lead to a variety of possible translations, such as the following:

> *For my dear companion, vowing to be well-deserving, E. Ver.*
> *As here revealed, praying to earn your friendship, E. Ver.*
> *Devoutly promising to be a well-deserving father, E. Ver.*

Bond concludes that regardless of how the translation is specified, one conclusion seems unavoidable . . . de Vere is giving his personal signature to the Dedication. In other words, *he wrote the Dedication himself.*

This conclusion is of course consistent with the first message we found in the Dedication: *THESE SONNETS ALL BY EVER* (with its possible qualifier *THE FORTH*).

It is certainly surprising to find a Latin sentence embedded in the Dedication, but can we somehow estimate just how surprising? We address this question in Appendix C. Ignoring the meanings of the words, and ignoring the word *EMERITER* (which is broken into two parts), we find that the probability of finding by chance a cluster of three Latin words of any meanings whatever, of lengths 3 letters, 4 letters, and 5 letters, comprising the letters we actually find in the Dedication, to be $4\,10^{-7}$ (DOB = −64).

If we allow for the possibility that the words might have been in a different order, this estimate would be increased by a factor of 20. If we were to consider that the cluster might have been located somewhere else in the grid, or in a different grid, the probability would again be increased. However, as Bond points out, the 12 × 12 grid is special, and the location of the cluster in that grid (on the left-hand edge and mid-height) also seems special, so it is not clear that one should consider arbitrary grid sizes and arbitrary possible locations in the grid. The tests that we carry out in Appendix C seem to confirm that this Latin sentence was intentionally built into the Dedication.

This of course leads to the question—who might have conceived of this sentence, in Latin, concealed as a cryptogram in the Dedication? Who had the motive, and who had the means? Thorpe may have had the means (knowledge of Latin), but did he have a motive? Did William Shakspere of Stratford-upon-Avon have either the motive or the means? These are some of the new questions that are posed by these investigations.

9. THE SHAKESPEARE MONUMENT AND INSCRIPTION

Figure 5 shows a picture of the monument to Shakespeare as it appears today in the Holy Trinity Church at Stratford-upon-Avon. Below the effigy is an inscription on a black background, as shown in Figure 22. It reads

IVDICIO PYLIVM, GENIO SOCRATEM, ARTE MARONEM,
TERRA TEGIT, POPVLVS MAERET, OLYMPVS HABET

STAY PASSENGER, WHY GOEST THOV BY SO FAST,
READ IF THOV CANST, WHOM ENVIOVS DEATH HATH PLAST,
WITH IN THIS MONVMENT SHAKSPEARE: WITH WHOME,
QVICK NATVRE DIDE. WHOSE NAME DOTH DECK Y^s TOMBE.
FAR MORE, THEN COST: SIEH ALL Y^T HE HATH WRITT,
LEAVES LIVING ART, BVT PAGE, TO SERVE HIS WITT.

OBIT ANO DO 1616
AETATIS 53 DIE 23 AP

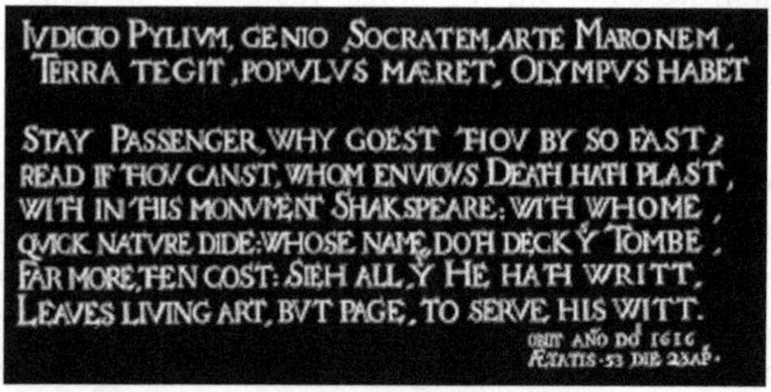

Figure 22. The inscription below the Shakespeare effigy monument in Holy Trinity Church (monument shown in Figure 5).

The following discussion of the inscription is based on the analysis of David Roper (2008, 2010). Roper (see Figure 23), born in London in 1938, lived with his grandparents in Lambeth, in London, following the death of his parents during the Blitz. Roper studied mathematics, statistics, and philosophy at the newly created Open University. After studying education for his postgraduate qualifications at Kingston University, he joined the staff at Reigate Grammar School, and later was Head of the Mathematics Department at Northwood College.

Roper's doubts about the Shakespeare Authorship began very early when, at the age of ten, on seeing one of the "Shakspeare" signatures, he recognized its similarity to the attempts made by young children when first practicing their name in cursive (joined-up) handwriting.

In 1988, following the publication of Charlton Ogburn's book (1988), Roper saw a television program in which Ogburn drew attention to the puzzle of the Shakespeare monument in the Holy Trinity Church, which is the subject of this section.

Figure 23. David Roper

Roper's attention was drawn to the curious phrase QUICK NATURE DIED. His knowledge of Latin led him to read *quick nature* as *velocium rerum*. This led Roper to consider this word-combination after all syllables after the first syllable in each word has "died" (a procedure familiar in crossword puzzles today). This train of thought led to *VElocium RErum*, i.e. *VE RE*, which got his attention. An elaboration of this theme led him to consider the Latin sentence *summa de velocium rerum* which, after it has "died," leads to SUM-ma DE VElocium RErum. *Sum De Vere* is Latin for *I am de Vere*.

Roper communicated this discovery to Lord Vere, who had written the Foreword to Ogburn's book. Lord Vere, in his response, invited Roper to join the De Vere Society, where Roper met John Rollett. Roper and Rollett carried out a lively correspondence for many years.

In this section, we review Roper's analysis of the inscription on the Shakespeare monument. Roper noted a number of peculiarities in this inscription, which strengthened his suspicion that the inscription contains something in code. He noted, specifically, the following seven peculiarities (Roper, 2008, 2017):

WHOM In line 2 is spelled differently from WHOME in line 3.
THIS in line 3 is written in full, but in line 4 it is abbreviated to Ys.
THAT is abbreviated to Yt in line 5.
The words, SHAKSPEARE MONVMENT, have been inverted in
    line 3 to read, MONVMENT SHAKSPEARE.
The name SHAKESPEARE has been spelled SHAKSPEARE.
The German word SIEH has been used in line 5 instead of SEE.
As WRITT, the word WRIT appears with an additional 'T'.

So many peculiarities led Roper to strongly suspect that the text conceals an encrypted message. This suspicion was reinforced by the fact that, as pointed out by Ogburn (1988, p. 157), "The Stratford monument, though wordy, cites no biographical fact about the deceased whatever."

Roper found that there is indeed a hidden message in the inscription on the monument. As was the case for the cryptograms in the Dedication of the Sonnets, the message has been encrypted by the ELS (equidistant letter sequencing) procedure—not a simple procedure to use. As the distinguished cryptographer David Kahn points out, *The method's chief defect, of course, is that awkwardness of phrasing may betray the very secret that that phrasing should guard: the existence of a hidden message* (Kahn, 1996, p. 144). Words that conceal a hidden meaning need to be chosen to accommodate the encrypted phrase or sentence without attracting suspicion.

One procedure for finding a hidden message involves guessing a probable word or name, which is referred to by cryptographers as a "crib," and examining its consequences. Roper considered the possible names "Bacon" and "Marlowe", but these proved not to be fruitful. He then considered the name "Edward de Vere," which proved to be the key to cracking the code. It led to an ELS sequencing of 34, which leads (as shown in Figure 24), to the message SO TEST HIM, I VOW HE IS E DE VERE AS HE, SHAKSPEARE: ME I. B. The letters I B are the initials of Ben Jonson (in reverse order, and using the letter I in place of J, which was not used in the Latin alphabet). A cryptographer would find it significant that the decrypted sentence occurs in three clusters.

Bruce Spittle, a New-Zealander and another Shakespeare scholar, noted that the second row of the inscription is indented. That row has 34 cells, suggesting that the author of the inscription was offering that number as a helpful hint (a "crib") to anyone searching the inscription for a possible hidden message.

Following Roper and Spittle, it seems reasonable to conclude that the message highlighted in Figure 24 was deliberately encoded in the 220 letters of the statement on the monument. If we paraphrase the message to clarify its significance, we might rewrite it as **I, Ben Jonson, vow that the works of Shakespeare were written by E de Vere.**

S	T	A	Y	P	A	S	S	E	N	G	E	R	W	H	Y	G	O	E	S	T	T	H	O	V	B	Y	S	O	F	A	S	T	R
E	A	D	I	F	T	H	O	U	C	A	N	S	T	W	H	O	M	E	N	V	I	O	V	S	D	E	A	T	H	H	A	T	H
P	L	A	S	T	W	I	T	H	I	N	T	H	I	S	M	O	N	V	M	E	N	T	S	H	A	K	S	P	E	A	R	E	W
I	T	H	W	H	O	M	E	Q	V	I	C	K	N	A	T	V	R	E	D	I	D	E	W	H	O	S	E	N	A	M	E	D	O
T	H	D	E	C	K	Y	S	T	O	M	B	E	F	A	R	M	O	R	E	T	H	E	N	C	O	S	T	S	I	E	H	A	L
L	Y	T	H	E	H	A	T	H	W	R	I	T	T	L	E	A	V	E	S	L	I	V	I	N	G	A	R	T	B	V	T	P	A
G	E	T	O	S	E	R	V	E	H	I	S	W	I	T	T																		

Figure 24. This is the Cardano Grid constructed by ELS (equidistant letter sequencing) for a spacing of 34, from the inscription on the Shakespeare Monument, using the crib "Edward de Vere."

The two lines of Latin look impressive to anyone who is not a Latin scholar. However, as Roper explains, they make little sense to anyone who is familiar with the language and with the personages referred to in the inscription.

We examine this cryptogram from a statistical viewpoint in Appendix D. That analysis leads to the conclusion that there is a probability of only 0.0002 (DOB = –37) of finding by chance a message containing the sequence EVERE in the inscription, when it is organized in a grid with a spacing of 34 cells.

10. DISCUSSION AND SUMMARY

We began by considering just two candidates for the role of the Great Author known as *Shakespeare*: the orthodox candidate, *William Shakspere* of Stratford-upon-Avon, and the strongest alternative candidate, *Edward de Vere, 17th Earl of Oxford*. A consequence of restricting the options in this way is that evidence *for* one candidate is evidence *against* the other—and vice versa.

A few of the basic concerns are listed in Table 3, where we compare skills and knowledge that we might expect to have been in the possession of the Great Author with what we learn about the two principal contenders, Shakspere and Oxford. We see that Oxford fares well, but Shakspere fares poorly.

In Section 4, we summarized an analysis of Diana Price's *Chart of Literary Paper Trails* (Price, 2012), which compares what we know of Shakspere related to the profession of writer with what is known of 24 known writers who lived in England at the same time as Shakspere. We found a significant mismatch that could have occurred by chance with an estimated probability of only 1 chance in 100,000 (DOB = –50).

TABLE 3
Properties Expected of the Author of the Works of Shakespeare,
Compared with Known Properties of William Shakspere and the 17th Earl of Oxford

	Shakespeare	Shakspere	Oxford
Evidence of:			
Extensive education	Yes	No	Yes
Superior handwriting	Yes	No	Yes
Extensive legal knowledge	Yes	No	Yes
Books, possession of	Yes	No	Yes
Travel, experience of	Yes	No	Yes
Foreign languages, knowledge of	Yes	No	Yes
Familiarity with nobility	Yes	No	Yes

Yes = evidence present; No = no evidence.

In Sections 5 and 6, we turned to an investigation of the *Dedication* of *Shake-speare's Sonnets*, reviewing analyses previously carried out by John Rollett and Jonathan Bond. We found that the Dedication contains several hidden messages. The first, in **The Dedication of the Sonnets . . .**, was a simple, unequivocal statement: THESE SONNETS ALL BY EVER THE FORTH. This message was obtained by selecting the 6th word, then the 2nd word of the remainder, then the 4th work of the remainder, and so on. We noted that the sequence 6 – 2 – 4 matches the number of letters in the name Edward de Vere. This message must have been deliberately built into the Dedication. *EVER* may obviously be read as a short representation of *Edward de Vere (E VERE)*. According to Rush (2016), de Vere often used *EVER* as his "signature."

However, the significance of *THE FORTH* remains obscure. Its interpretation as *THE FOURTH* remains unconvincing: It may have had special significance only for the intended recipient of the Sonnets and Dedication, most likely the *Fair Youth* (who is widely believed to have been Southampton). Hence the true significance of *THE FORTH* may become clear only when we have a complete understanding of the relationship between Oxford and Southampton.

In Section 8, in our analysis of the hidden Latin message PRO PARE VOTIS EMERITER that was discovered by Jonathan Bond, we noted the possible interpretations proposed by Bond. One of these is *Devoutly promising to be a well-deserving father*. We may note that Bond's proposal receives some support from a possible reinterpretation of the word FORTH in the decryption THESE SONNETS ALL BY EVER THE FORTH, which we discussed in Section 6. As we mentioned in that section, the word FORTH has to have some significance but that significance continues to elude Shakespeare scholars.

Bond's suggestion was that the Dutch for "the fourth" is "de vierde," which is phonetically close to "de Vere." However, if we move one letter by just two places, we arrive at the anagram THESE SONNETS ALL BY EVER THE FOTHR, which when read aloud is very close to THESE SONNETS ALL BY EVER THE FOTHER. Since FOTHER was an alternative spelling of FATHER in Elizabethan writing, this offers some tenuous support to Bond's conjecture that the Latin message might be interpreted as *Devoutly promising to be a well-deserving father.*

As we mentioned in Section 6, it is possible that—400 years later—the true significance of THE FORTH, if there is one, may never be known. If it was a reference to an in-joke between the author and the dedicatee, its significance will remain forever a matter of conjecture.

From Section 7 on, we concentrated on a search for messages that had been encrypted by the ELS (Equidistant Letter Sequencing) procedure. The first discovery (by Rollett) was found in a grid with 15 letters in each row—the name HENRY. We estimated the probability that the name HENRY might have occurred by chance in one of a wide range of possible grids to be 0.002 (DOB = −27). This discovery led Rollett to carry out a close examination of a wide range of possible grids. In the grid with 8 letters in each row, Rollett found the letter-groups WR, IOTH, and ESLEY which, when combined, spell WRIOTHESLEY, the surname of Henry Wriothesley, the Third Earl of Southampton. As we have noted earlier, Southampton is the only person whose name appears in any of the Shakespeare oeuvres (namely in the dedications of the long poems *Venus and Adonis* and *Rape of Lucrece*). He is recognized as the leading candidate for the identity of the *Fair Youth* of the Sonnets.

We estimated the probability of finding the name *WRIOTHESLEY*, broken up into either 2 or 3 parts, anywhere in a wide range of grids, to be $7 \cdot 10^{-6}$ (DOB = −52). Hence the probability that the full name *HENRY WRIOTHESLEY* might have appeared by chance, in a combination of 3 or 4 grids, is approximately 10^{-8}, i.e. one part in one hundred million (DOB = −80). Combining this estimate with the conservative estimate (in Section 8) of 7×10^{-8} of the appearance of the (Latin) cluster *PRO PARE VOTIS EMERITER*, we arrive at an estimate of about 10^{-15} (DOB = −150) of finding in the Dedication the above two sequences. How can we visualize the significance of such a small number? We can follow the example of John Rollett (1997a, 1997b).

Rollett estimated that one ton of sugar contains approximately one billion (10^9) grains. Hence the probability of finding the above two sequences by chance is approximately the same as extracting, by chance, a specific grain (say one that had been colored red) out of one million tons of sugar! It would take a cubic container with a side dimension of order 100 meters to hold one million tons of sugar.

The statistical evidence is overwhelming that the Dedication is intended to inform us that the great writer we know as Shakespeare was not William Shakspere of Stratford-upon-Avon, but Edward de Vere, Earl of Oxford. The Dedication is rather like a message inserted in a bottle thrown into the ocean 400 years ago, and retrieved only in the last century—not by accident but by the brilliance and perseverance of Rollett and Bond.

To the discoveries of Rollett and Bond, we must add the discovery of Roper (also a mathematician), which we discussed in Section 9. The strange inscription on the Shakespeare Monument itself contains a hidden message from the pen of Ben Jonson, confirming the true authorship of the works of "Shakespeare." His composition does not have the impressive elegance that we found in the Dedication of the Sonnets. However, to be fair, Jonson may have had only a few days in which to compose his cryptogram, whereas de Vere may have spent months on his incredible composition.

As often happens in solving one problem, we now face many new ones, among them:

Why and when did Oxford begin using the pen name Shakespeare?
Why and when did Oxford stop using the pen name Shakespeare?
Who decided that the identity of the Great Author should be fastened
    upon William Shakspere of Stratford-Upon-Avon?
Was Shakspere paid to be a party to that deception?
Was Ben Jonson a party to that deception?
What were the relationships among Oxford, Southampton, and the
    Queen?

Why did the normally tight-fisted Queen grant Oxford an annuity of
£1,000?

Why did King James I continue the annuity after the Queen's death?

Who promoted and who financed the First Folio?

What was Ben Jonson's role in the production of the First Folio?

Why did the First Folio not include Shakespeare's poems?

What were the goals of Essex and Southampton in their disastrous
"rebellion"?

Who would have become king, had they succeeded?

Why was Southampton not executed, as was Essex?

Why did James restore to Southampton all of his titles and possessions?

Finally, we comment very briefly on the possible origin of the
Dedication. We have noted that Duncan-Jones (1997), Booth (2000),
and Vendler (1997), who all comment extensively on the Sonnets,
attributed the Dedication to the publisher Thomas Thorpe. In view
of our analyses of the hidden content of the Dedication, this option
must be abandoned. Apart from the highly significant content of the
Dedication that has nothing to do with the interests of Thorpe, we
must recognize that it would take weeks—at least—of concentrated
effort to generate such an intricate array of cryptograms. Whoever
generated the Dedication had extraordinary facility with both English
and Latin, knowledge of cryptography, probably weeks or more that
could be devoted to the task, and a consuming incentive to share
highly sensitive information with whomever might have the time,
skill, and motivation to devote to a searching but tedious study of the
Dedication. That would be a lot to ask of Thomas Thorpe.

Bond has considered this question as part of his analysis of the
hidden content of the Dedication, and concluded (Bond, 2009, p. 81)
that *de Vere gave his personal signature to the Dedication—in other words,
he wrote the Dedication himself.* We see no reason to disagree.

There is one consideration that might give one pause—the Dedication
refers to *Our Ever-Living Poet.* The term *ever-living* was used only in relation
to someone who was dead. How can we resolve this paradox?

First, we can note that in the Sonnets (see Sonnet 17, for instance)
de Vere was writing not just for his immediate contemporaries but for
readers yet to come. Second, to reveal his identity, he had to find a way

to include in the Dedication the word EVER.

In view of the extraordinary goal that de Vere had set himself, surely we can cut him some slack and allow him to use a term—*ever-living*—that may not have been 100 percent appropriate.

ACKNOWLEDGMENTS

The authors acknowledge with thanks their communications with Jonathan Bond, Katherine Chiljan, John Hamill, Martin Hellman, Diana Price, David Roper, Jeffrey Scargle, and Jessica Utts. We owe special thanks to Jonathan Bond, Diana Price, and David Roper for allowing us to quote from their publications. Historical photos were reprinted from open archives at the Bodleian Library at Oxford University and from Wikipedia. Researcher photos were obtained from the subjects. Table 1 was researched and created by Kathleen Erickson. All other tables were compiled by Peter Sturrock.

REFERENCES

Anderson, M. (2005). *"Shakespeare" by another name: The life of Edward de Vere, Earl of Oxford, the man who was Shakespeare*. Gotham Books.

Bacon, D. (1856, January). William Shakespeare and his plays: An inquiry concerning them. *Putnam's Magazine of American Literature, Science, and Art, VI*(37), 1–19.

Bianchi, J. S. (2018). Twins separated by birth?: A cultural and genealogical investigation of two identities set in stone. *Shakespeare Oxford Newsletter, 54*(4), 23.

Bond, J. (2009). *The De Vere code: Proof of the true author of Shake-Speares Sonnets*. Real Press.

Booth, S. (2000). *Shakespeare's Sonnets*. Yale University Press.

Chiljan, K. (2011). *Shakespeare suppressed: The uncensored truth about Shakespeare and his works, A book of evidence and explanation*. Faire Editions.

Clark, E. L. T. (1931/1972). *Hidden allusions in Shakespeare's plays: A study of the Oxford theory based on the records of early Court Revels and personalities of the times*. William Farquhar Payson [printed by Stratford Press].

Duncan-Jones, K. (1997). *Shakespeare's Sonnets*. Nelson and Sons.

Friedman, W. F., & Friedman, E. S. (1957). *The Shakespearean ciphers examined: An analysis of cryptographic systems used as evidence that some author other than William Shakespeare wrote the plays commonly attributed to him*. Cambridge University Press.

Gallup, E. W. (1910). *Concerning the bi-literal cypher of Francis Bacon discovered in his works*. Howard.

Hotson, J. L. (1964). *Mr. W. H.* Rupert Hart-Davis.

Jaynes, E. T. (2003). *Probability theory: The logic of science* (G. Larry Bretthorst, Ed.). Cambridge University Press.

Kahn, D. (1996). *The code-breakers: The comprehensive history of secret communication from ancient times to the Internet*. Charles Scribners' Sons.

Looney, J. T. (1920). *"Shakespeare" identified in Edward de Vere, the Seventeenth Earl of Oxford*. Cecil Palmer; Frederick A. Stokes.

Michell, J. (1996). *Who wrote Shakspeare?* Thames and Hudson.

Nelson, A. H. (2003). Monstrous adversary: The life of Edward de Vere, 17th Earl of Oxford. *Liverpool English Texts and Studies, 40*. Liverpool University Press.

Nolen, S. (2010). *Shakepeare's face: Unraveling the legend and history of Shakespeare's mysterious portrait*. Simon and Schuster.

Ogburn, C., Jr. (1988). *The mystery of William Shakespeare*. Sphere Books.

Pointon, A. J. (2011). *The man who was never Shakespeare: The theft of William Shakspeare's identity*. ParaPress.

Price, D. (2012). *Shakespeare's unorthodox biography: New evidence of an authorship problem*. Greenwood.

Rollett, J. M. (1997a). The Dedication to Shakespeare's Sonnets, Part 1: Mr W. H. revealed at last. *Elizabethan Review, 5*(2), 93–106.

Rollett, J. M. (1997b). The Dedication to Shakespeare's Sonnets, Part 2: These. sonnets. all. by. . . . *Elizabethan Review, 5*(2), 107–122.

Rollett, J. M. (1999). Secrets of the Dedication of Shakespeare's Sonnets. *The Oxfordian, 2*, 60–75.

Rollett, J. M. (2004). *Secrets of the Dedication of Shakespeare's Sonnets.* Parapress.

Rollett, J. M. (2015). *William Stanley as Shakespeare: Evidence of authorship by the Sixth Earl of Derby*. McFarland.

Roper, D. L. (2008). *Proving Shakespeare: The looming identity crisis*. Orvid Books.

Roper, D. L. (2010). *Shakespeare: To be or not to be?* Orvid Books.

Roper, D. L. (2017). William Shakespeare: A study of the poet and five famous contemporaries who between them used the rune ciphers to reveal his true identity. *Journal of Scientific Exploration 31*(4), 625–676.

Roper, D. L. (2018). *How science proved Edward de Vere was Shakespeare*. First Proofs.

Rush, P. (2016), *Hidden in plain sight: The true history revealed in Shake-Speares Sonnets*. Read-Deal Publications.

Sanders portrait. (2020, June). Sanders portrait of Shakespeare. In Wikipedia. https://en.wikipedia.org/wiki/Sanders_portrait

Schoenbaum, S. (1970). *Shakespeare's lives*. Clarendon Press.

Story, T. (2016). *The Shakespeare fraud: The politics behind the pen* (William Boyle, Ed.). Forever Press.

Sturrock, P. A. (1994). Applied scientific inference. *Journal of Scientific Exploration, 8*(4), 491–508.

Sturrock, P. A. (2008). Shakespeare: The authorship question, A Bayesian approach. *Journal of Scientific Exploration, 22*(4), 529–537.

Sturrock, P. A. (2013). *AKA Shakspeare: A scientific approach to the authorship question.* Exoscience.

Vendler, H. H. (1997). *The Art of Shakespeare's Sonnets.* Harvard University Press.

Wells, S. (1970). *The Oxford Shakespeare: The complete works* (2nd ed.). Oxford.

APPENDICES

Appendix A. Concerning the Name HENRY

We here assess the significance of finding the word HENRY.

TABLE A-1
Analyzing the Word HENRY

H	10	144	0.0694
E	23	143	0.1608
N	13	142	0.0915
R	9	141	0.0638
Y	1	140	0.0071

Column 1: Letter. Column 2: Number of such letters available.
Column 3: Number of letters left to choose from. Column 4:
Probability of finding that letter in that cell.

We begin by considering the letter H. We see from column 2 that there are 10 letter H's in the text. We see from column 3 that there are 144 cells in the array. Hence, as shown in Column 4, the probability of finding a letter H in a cell chosen at random is 10/144, i.e. 0.0694.

When we come to letter E, we find that there are 23 letter E's in the remaining text, which is now reduced to 143 letters. Hence, as shown in Column 4, there is a probability of 23/143, i.e. 0.1608 of finding an E in a cell chosen at random from the 143 available cells.

We proceed similarly for the letters N, R, and Y, using the remaining 3 rows of the table.

We can now calculate the probability of finding the letters H,E,N,R,Y in a set of 5 cells chosen at random by forming the product of these 5 probabilities, which is found to be $4.63 \cdot 10^{-7}$.

However, we now need to take account of all the ways that the author could have selected a sequence of 5 cells. There are 10 rows in that grid, so one can fit 5 letters as a sequence in a given column in 6 ways—starting with the top cell, the second cell, etc., down to the sixth cell. Hence the probability of finding HENRY in any given column of 10 cells is $6 \cdot 4.63 \cdot 10^{-7}$, which is $2.8 \cdot 10^{-6}$. But remember that we are willing to have this word read from top to bottom or from bottom to top, which increases the probability by a factor of 2, giving us a probability of 5.6 10^{-5}.

We next take account of the fact that the grid we are considering has 15 columns, so the probability of finding HENRY (either up or down) somewhere in that grid is 15*5.6*10^{-5}, which is 8.4 10^{-5}.

This is the estimate for a single grid with 10 rows (the one that shows the name HENRY). However, we could fit the name HENRY into a grid with only 5 rows. If we look for grids with 5 rows or more and 5 columns or more, we find that there are 21 such grids.

Taking this factor into account, we find that the probability of finding the name HENRY by chance somewhere in one of the possible grids is 21*8.4*10^{-5}, which is approximately 1.9 10^{-4} (DOB = −27).

Appendix B. Concerning the Name WRIOTHESLEY

We now assess the significance of finding the word WRIOTHESLEY.

TABLE B-1
Analyzing the Word WRIOTHESLEY

W	4	144	0.0278
R	9	143	0.0629
I	14	142	0.0986
O	8	141	0.0567
T	17	140	0.1214
H	10	139	0.0719
E	23	138	0.1667
S	10	137	0.0730
L	6	136	0.0441
E	22	135	0.1630
Y	1	134	0.0075

We first count the number of times each letter occurs in the text.
Column 1: Letter. Column 2: Number of such letters available.
Column 3: Number of letters left to choose from. Column 4:
Probability of finding that letter in that cell.

The product of those probabilities is found to be 5.58 10^{-14}. This is the probability of finding the sequence WRIOTHESLEY in 11 cells by chance.

We next consider the possibility that the letters WRIOTHESLEY might have been organized in just two columns, with results shown in Table B-2. The columns contain the following: column 1: the number of letters in one column; column 2: the number of letters in the other column (these two numbers must sum to 11); column 3: the number of ways one can arrange the letters in column 1 in a column with just 8 cells (this is 9 minus the number); column 4: the number of ways one can arrange the letters in column 2 in a column with just 8 cells (this is 9 minus the number); column 5: the product of the numbers in columns 3 and 4.

TABLE B-2

3	8	6	1	6
4	7	5	2	10
5	6	4	3	12
6	5	3	4	12
7	4	2	5	10
8	3	1	6	6

The number of ways that one can arrange the 9 letters in 2 columns is the sum of the numbers in column 5. This is found to be 56. However, since one may need to read a sequence either from top to bottom or from bottom to top, we must multiply this number by 4, to obtain 224. There are $^{18}C_2$, i.e. 153, ways of selecting two columns out of 18. With this factor, we find that there are 34,272 ways of entering 11 letters in the grid, using only 2 columns of the grid.

We now repeat these calculations on the assumption that the letters are distributed in 3 columns. Now, restricting the options to 2 or more letters per column, the possible arrangements are found to be (Table B-3):

TABLE B-3

2	2	7
2	3	6
2	4	5
2	5	4
2	6	3
2	7	2
3	2	6
3	3	5
3	4	4
3	5	3
3	6	2
4	2	5
4	3	4
4	4	3
4	5	2
5	2	4
5	3	3
5	4	2
6	2	3
6	3	2
7	2	2

We now proceed as before, calculating the number of ways of entering 11 letters in 3 columns of 8 cells each as follows (Table B-4):

TABLE B-4

2	2	7	7	7	2	98
2	3	6	7	6	3	126
2	4	5	7	5	4	140
2	5	4	7	4	5	140
2	6	3	7	3	6	126
2	7	2	7	2	7	98
3	2	6	6	7	3	126
3	3	5	6	6	4	144
3	4	4	6	5	5	150
3	5	3	6	4	6	144
3	6	2	6	3	7	126
4	2	5	5	7	4	140
4	3	4	5	6	5	150
4	4	3	5	5	6	150
4	5	2	5	4	7	140
5	2	4	4	7	5	140
5	3	3	4	6	6	144
5	4	2	4	5	7	140
6	2	3	3	7	6	126
6	3	2	3	6	7	126
7	2	2	2	7	7	98

In Table B-4, columns 1 to 3 list the number of cells occupied by letters. Column 4 lists the number of ways of arranging the number of letters listed in column 1 in 8 lines, etc. Column 7 lists the products of the numbers in columns 4 to 6. The total number of ways of arranging 11 letters in 3 columns is the sum of the numbers listed in column 7, which is found to be 2,772.

Allowing for the up–down ambiguities (a factor of 8), this becomes 22,176. The number of ways of selecting 3 columns out of 18 is $^{18}C_3$, i.e. 816. With this factor, the number of options becomes 18,095,616. If we add the number for the two-column case, we get 18,129,888. Combining this with the basic factor of 5.58×10^{-14}, we estimate the probability of finding the name WRIOTHESLEY by chance in the 18 x 8 grid to be 1×10^{-6}.

However, there are six other "perfect grids" that have 6 or more columns and 6 or more lines: 24 x 6; 16 x 9; 12 x 12; 9 x 16; 8 x 18; and 6 x 24. Assuming that analyses of these grids give similar results, we estimate the probability of finding the name WRIOTHESLEY in one of the 7 grids to be 7×10^{-6} (DOB = −52).

Appendix C. Concerning the Phrase "Pro Pare Votis Emeriter"

We here assess the significance of finding the phrase PRO PARE VOTIS EMERITER. Since the word EMERITER is broken up into three pieces, we shall ignore that word. Table C-1 shows the number of times each of these letters occurs in the text. The columns show: 1: Letter; 2: Number of times each letter occurs in the text.

TABLE C-1
Concerning the Phrase PRO PARE VOTIS EMERITER

A	5
E	23
I	14
M	2
O	8
P	4
R	9
S	10
T	17
V	6

We first consider the word PRO. The probability of finding the letters P, R, O by chance in a specified group of 3 cells is:

$$P(PRO) = (4/144) \times (9/143) \times (8/142) = 9.85 \ 10^{-5}$$

However, from examination of a Latin dictionary, I have estimated that there are 92 3-letter words in Latin, so I estimate that the probability of finding the word PRO or any other 3-letter word in the Dedication is

$$P(\text{word like PRO}) = 92 \times 9.85 \ 10^{-5} = 0.0091$$

We next consider the word PARE. Noting that we have already used the letters P,R,O, we find the probability of finding by chance the letters P,A,R,E in a specified group of 4 cells is:

$$P(PARE) = (3/141) \times (5/140) \times (7/139) \times (23/138) = 6.4 \ 10^{-6}$$

But we estimate that there are 410 4-letter words in Latin, from which we estimate that

$$P(\text{word like PARE}) = 2.6 \times 10^{-3}$$

We now consider the word VOTIS. Proceeding as before, we estimate

$$P(VOTIS) = (6/137) \times (7/136) \times (17/135) \times (14/134) \times (10/133) = 2.2 \ 10^{-6}$$

However, we estimate that there are 1,150 5-letter words in Latin, from which we estimate that

$$P(\text{word like VOTIS}) = 0.0026$$

Hence the probability of finding by chance, in specified locations, one word like PRO, one word like PARE, and one word like VOTIS is

$$P(\text{words like PRO PARE VOTIS}) = 6.2 \ 10^{-8}$$

We may now consider the possible locations of this group of 3 words. As we see from Figure 21, PRO, PARE, and VOTIS are found in the first three columns, so it is reasonable to leave them in those columns. The vertical spacing is so arranged that one can read the words PRO and EVER in rows 7 and 8, respectively. Hence it seems reasonable to leave the relative vertical locations unchanged.

With these stipulations, we find that there are 6 possible vertical locations that leave unchanged the relative locations of PRO, PARE, and VOTIS. Hence we finally arrive at our estimate of the probability that three words of the same lengths as PRO, PARE, and VOTIS could have appeared by chance in a group similar to the actual location is given by

$$P(\text{final}) = 6 \times 6.2 \times 10^{-8} = 4 \ 10^{-7} \ (DOB = -64)$$

Taking account of the word EMERITER would reduce this probability.

Appendix D. Concerning the Inscription on the Monument

We can obtain a conservative estimate of the significance of the sentence

SO TEST HIM, I VOW HE IS E DE VERE AS HE, SHAKESPEARE: ME I.B.

by estimating the probability of finding the letter sequence EVERE in the inscription, given the letter count (shown in Table D-1).

TABLE D-1
Word Breakdown of the Inscription on the Monument

E	25
R	9
V	11
TOTAL	220

We first find the probability of finding the sequence EVERE in a sequence of five cells. The number of occurrences of E, R, and V and the total number of letters in the inscription are shown in Table D-1.

The probability of finding the first letter E in a given cell is 25/220.

The probability of finding the letter V in a remaining cell is 11/219.

The probability of finding the second letter E in a remaining cell is 24/218.

The probability of finding the letter R in a remaining cell is 9/217.

The probability of finding the third letter E in a remaining cell is 23/216.

Hence the probability of finding the sequence EVERE in a given sequence of 5 cells is $2.8 \ 10^{-6}$.

We see from Figure 24 that, when the text is arranged in a grid with 34 columns, there are 16 columns of length 7 and 18 of length 6.

A column of length 6 has 2 possible sequences of 5 cells.

A column of length 7 has 3 possible sequences of 5 cells.

Hence the number of possible sequences of 5 cells is $16 \times 3 + 18 \times 2$, i.e. 84. Hence the probability of finding the sequence EVERE in a grid of the shape shown in Figure 24, if the 220 letters are distributed at random in the 220 cells, is given by $84 \times 2.8 \times 10^{-6}$, i.e. approximately $2 \ 10^{-4} \ (DOB = -36)$.

TABLE 1
A Timeline of Relevant Events and Publications

The purpose of this Timeline table is to summarize, very briefly, relevant events for which the dates are known, and to display them in a chart that enables one to see at a glance what is known to have been happening in the life of Shakspere, etc., and the relationship—or lack of relationship—of these events to the publications of the plays and poems that are attributed to "Shakespeare."

Column 1	Date
Column 2	Publications
Column 3	Events in the life of William Shakspere
Column 4	Events in the life of Edward de Vere
Column 5	Other relevant events

Reference Key: A = Anderson, 2005. Ch = Chiljan, 2011. C = Clark, 1931. L = Looney, 1920. N = Nelson, 2003. P = Pointon, 2011. Pr = Price, 2012. R = Roper, 2018. S = Story, 2016.

Summary of events before the births of Shakspere and Oxford:

1215 June 15: Edward de Vere's 12th great grandfather, Robert de Vere, the 3rd Earl of Oxford, was one of a group of barons who forced King John I to sign the Magna Carta (C328).

1256 Fuller's *Worthies* said Aubrey de Vere was the greatest scholar of the age.

1386 The 9th Earl of Oxford was Duke of Ireland and a consort of King Richard II (A1).

1415 The 11th Earl of Oxford was a commander in the Agincourt battle serving Henry V (A1).

1516 February: Henry VIII and first wife Catherine of Aragon, after 5 miscarriages and infant deaths, produced a daughter, Mary Tudor (S14).

1519 June: Henry VIII has a bastard son with Elizabeth Blount and names him Henry FitzRoy (S14).

1521 September: William Cecil (later Lord Burleigh) is born to an innkeeper (S21).

1525 June: Henry FitzRoy made Duke of Richmond and Knight of the Garter (S14).

1533 September: Wanting a legitimate son, Henry VIII makes England Protestant in order to divorce Catherine and marry Anne Boleyn, who bears the future Queen Elizabeth I (S14).)

1535 William Cecil enters Cambridge University, a hotbed of Protestant thinking (S21).

1536 May: Still with no son, Henry VIII executes Anne Boleyn and marries Jane Seymour (S14). July: Henry VIII's 2nd Act of Succession gives him the right to decide his successor and delegitimizes Mary and Elizabeth (S15). July: Henry FitzRoy dies (S15).

1537 October: Edward VI born to Henry VIII and Jane Seymour, and he is next in line for the throne (S15). Jane dies shortly after the birth (S15).

1540 Henry VIII married Ann of Cleves but annulled it after 6 months (S15). Henry VIII married Catherine Howard whom he beheaded for being unfaithful (S15).

1541 January: William Cecil married Mary Cheke, daughter of a Cambridge professor (S21) who later is tutor to King Edward VI (S23). Mary bears him a son, Thomas, in 1542 and then dies in 1543 (S21).

1543 Henry VIII married Catherine Parr, who survived him (S15). July: Henry VIII's 3rd Act of Succession returns Mary and Elizabeth to the line of succession (they remain technically bastards) behind Edward (he is sickly) (S15).

1546 William Cecil married Mildred Cooke whose father was a leading supporter of the Reformation (S21).

1547 January: Henry VIII died and Edward VI became King at age 10 (S15).

1548 Edward VI became child king, his uncle the Duke of Somerset became his Protector, and William Cecil became the Protector's Secretary (S21).

TABLE 1 (continued)

YEAR	SHAKESPEARE	SHAKSPERE	OXFORD	OTHER
Year	Shakespeare's Work (play dates from *Court Revels* [C], Ch, et al.)	William Shakspere of Stratford-upon-Avon	Edward de Vere, 17th Earl of Oxford	Court Events and other Contemporary Events
1550			Born April 12 at Castle Hedingham (L) and named after Henry VIII's only son Edward VI, who sent a gilded chalice for the christening (A2). At this time, his father, the 16th Earl of Oxford, owned 300 castles and mansions (A1).	Somerset overthrown by John Dudley, Earl of Warwick, and Cecil arranged for Somerset's execution (S23). Cecil became Secretary of State to Dudley (S21). Girolano Cardano proposed a cryptographic method (derived from Cabbalist coding practices) later called Cardano Grilles, used in European diplomacy for centuries (R8).
1551		John Shagspere (William's father) at age 20 bought his early freedom from apprenticeship and set up shop as a glover (S84).		Cecil was knighted and given several estates (S21).
1552		John Shakspere was fined for keeping midden in Henley St. (P269).		
1553			June: 16th Earl of Oxford (father) signed letters nominating Lady Jane Grey as successor. July: 16th Earl of Oxford declared for Mary for Queen instead. Sept.: Joined Privy Council (N22).	Jan.: Death of King Edward VI at age 16. He had nominated Lady Jane Grey as his successor and she ascended the throne for 9 days (S15). July: Mary I asserted her Tudor blood-right and became Queen (S22,S16), and Lady Grey was executed (S16). Mary had married Philip II of Spain, made England Catholic again, reintroduced the Inquisition to England, and earned the name "Bloody Mary" (S16). William Cecil laid low during Mary's reign (S22).
1554–1562			Under tutelage of Thomas Smith at Ankerwicke N25).	

TABLE 1 (continued)

YEAR	SHAKESPEARE	SHAKSPERE	OXFORD	OTHER
1556		John Shakspere was named as a glover (P269). John appointed as Taster of Bread and Ale (P269). John sued Henry Field for 4.5 tons of barley (P269).	Future wife Anne Cecil born (L). Access to Elizabeth (soon to be queen).	Death of Queen Mary. Accession of Queen Elizabeth (L).
1557		John Shagspere married Mary Arden who had just inherited property (S84). He was given the title Official Ale Taster (S84).		
1558		John bought Henley St. property (P269). His wife Mary inherited her father's land at Wilmecote (P269).	Performed at Court. Student at Queen's College, Cambridge University, 1558–1559 (N23). Tutored by Thomas Fowle (who received 10 pounds per year), and whose anti-Catholic activities were suppressed by royal authority (N25).	Nov.: Queen Mary died after a false pregnancy (S22, S16). Queen Elizabeth ascended the throne at age 25. The 16th Earl of Oxford officiated as Lord Great Chamberlain. The Queen made England Protestant again (S16). She appointed William Cecil Principal Secretary to the Queen and a Privy Counselor (S22).
1559				Coronation of Queen Elizabeth. Jan.: House of Commons petitioned the Queen to marry, and her answer was that it was between herself and God (S17). June: Rumors that the Queen was with child or had already had children with Dudley (S17).
1560				Sept.: Dudley's wife died amid rumors he poisoned her to be free for the Queen (S517,S17).
1561		John Shagspere elected Chamberlain of the Borough of Stratford (S84, P269).	Queen Elizabeth visited Castle Hedingham for five days (A12, N29, S18). Edward was 11 years old.	July: New rumors that the Queen looked as if she had come out of childbed, by the Duchess of Suffolk (S18). Aug.: Queen visited the 16th Earl of Oxford, the highest-ranking nobleman in England and her Lord Chamberlain, who officiated at her coronation, for five days.

TABLE 1 (continued)

YEAR	SHAKESPEARE	SHAKSPERE	OXFORD	OTHER
1562	*Romeo and Juliet* alluded in the contemporary literature in 6 texts.		Contracted to marry into the Hastings family (potential heirs to the throne) (A15). Aug.: Father died (N30). Edward escorted by 140 men on horseback to become a Royal Ward at William Cecil's house (N35, S18). Cecil's library had texts in Latin, Greek, French, Italian, and Spanish (S27, A21). His wife Mildred Cooke was highly educated (S27). His Uncle Arthur Golding was his tutor (S27). He entered St. John's College, Cambridge University. His mother remarried. 32 of his 77 properties came under his control, on which he received 5% a year.	Cecil House, with a magnificent library and steps from the palace, was completed just in time for the beginning of Oxford's nine-year stay as ward of the Queen and Cecil (S22). Queen Elizabeth was deathly ill with smallpox (S18,A21). The Queen named Robert Dudley as potential lord protector of England, "Protector of the Realm," to everyone's shock (S18).
1563			Tutored by Anglo-Saxonist Laurence Nowell, who signed the Beowulf ms that was in Cecil House, the only known copy in the world (A23).	June: William Cecil's second son, Robert, was born, dropped on his head, and crippled. Nov.: In response to the Queen's smallpox scare, the House of Lords petitioned her to marry and produce an heir: She discontinued Parliament for 3 years, until she needed money (S18).
1564		Apr. 28: William was baptized at the Holy Trinity Church (S84, P269). His father became an Alderman (S84).	Received a Bachelor's degree from Cambridge University.	The Queen promoted Robert Dudley to Earl of Leicester.
1565	Shakespeare quoted from all 15 books of Golding's English translations of *Ovid's Metamorphosis* (A27), his most influential source after the Bible (A27).		His uncle and tutor Arthur Golding translated Ovid's *Metamorphosis*.	1565–1597 Golding's translation of *Ovid's Metamorphosis* was published in 7 editions.
1566			Received a Master's degree from Cambridge University at 16 yrs old (S25).	Marriage arrangement made for Anne Cecil (12 yrs old) and Philip Sidney (15 yrs old) (L).
1567			Admitted to Gray's Inn for legal studies (S25) with Philip Sidney (M33).	Lord Darnley, husband of Mary Queen of Scots, was murdered.

TABLE 1 (continued)

YEAR	SHAKESPEARE	SHAKSPERE	OXFORD	OTHER
1567	The first story in Paynton's *Palace of Pleasure* was an early version of *Timon Athens* (C41).		Injured servant Thomas Brincknell with a rapier during fencing practice, and Thomas died later that day (N47).	Queen Mary abdicated after her husband's murder scandal, and her son James became King of Scotland.
1568		John was made Bailiff of Stratford (P269).	His mother died (L).	Former Queen Mary fled to England, and was imprisoned by Queen Elizabeth.
1569		John applied for a coat of arms to make himself a gentleman, claiming his grandfather was a hero in the War of Roses and was granted land by Henry VII in 1485. The Herald's Office turned him down (S84). John sued a debtor who had purchased wool (P269).	Sought military service and was refused by the Queen (L, N54). His earliest surviving letter is from this year (N62). 74 letters and notes survive today in Oxford's handwriting, more than 50,000 words, with much creative use of spelling (N63).	June: 2nd Earl of Southampton entertained the Queen at his Titchfield castle (S38). Earls of Northumberland and Westmoreland led a Northern rebellion against the Elizabethan state. The French Ambassador reported to Paris that English nobles were convinced that the Queen would never marry and that William Cecil was seeking "other means" to solve the succession crisis (S25).
1570		John became Mayor of Stratford (S85). Prosecuted for usury and fined 40s (P269) for illegal dealing in wool and usury. He was charging 20–25% interest, while anything over 10% was illegal (S85).	Wrote short poems about a first-person nobleman-courtier horseman, a witty and persuasive speaker (Ch76), who was educated in law and rhetoric. These poems have echoes of Shakespeare's *A Lover's Complaint* and Robert Chester's *Love's Martyr: Or Rosalin's Complaint*. Recovered from an illness at an inn in Windsor. Joined the Earl of Sussex in military action to suppress a rebellion of Northern Earls in the English border counties and southern Scotland. Purchased the following books: A Geneva Bible (now in the Folger Library), a Chaucer, Plutarch's works in French and other books in French, two books in Italian, folio copies of Cicero and Plato (N53).	June: 2nd Earl of Southampton detained by Sheriff of London for Catholic sympathies and activities (S38) and in July sent to William More's home in Losley—in his custody until he conformed by joining in family prayers (S38). Dec.: 2nd Earl of Southampton submits and is released (S38). Pope Pius V declared Elizabeth's reign illegitimate, called for English Catholics to assassinate her (S19), excommunicated her (S19), and released English Catholics from their obligation to obey her (S19). July: William Cecil memo refuted charges that Oxford received money from his estates while he was in Italy. This is the only evidence that Oxford was in Italy before his 1575 trip there.

TABLE 1 (continued)

YEAR	SHAKESPEARE	SHAKSPERE	OXFORD	OTHER
1571		John was made Deputy Bailiff (P269).	Apr.: Turned 21 years old, left wardship with Cecil, and joined Parliament (L). He became a candidate for the Knight of the Garter, the highest order in England, and received 10 first-place votes (S26), but the Queen did not approve it (S30). His first cousin the Duke of Norfolk was imprisoned. July: Became engaged to 15-yr-old Anne Cecil, his former guardian Lord Burleigh's daughter. Dec.: Married Anne Cecil (S26), and made over Castle Hedingham and other properties to Lord Burleigh (which happened with other wards of Burleigh's) (L), a similar situation to Bertram in *All's Well That Ends Well* — marrying someone who grew up in the same household but who is socially inferior (L). The Queen attended the marriage (L). May: Oxford distinguished himself in jousting, winning over Howard (L, S26, A46), and received a book with diamonds on the cover from the Queen (C22, S26). His 3 best friends were cousin Thomas Howard, the most learned nobleman of his time, Charles Arundel; and Robert Southwell (all 10 yrs older) (N59). Listed as a Friend in a Catholic memo (Ridolfi Plot) (N68).	Feb.: William Cecil elevated to Baron (Lord) Burleigh and Chancellor of Cambridge University (L), and the marriage contract for Anne Cecil and Philip Sidney was cancelled (L). Feb.: Cecil, Lord Burleigh, made Knight of the Garter and Lord Treasurer, which made him the most powerful man in England (S26). the Duke of Norfolk was imprisoned for attempting to marry Mary, former Queen of Scots, and depose Elizabeth. Apr.: New law prohibited discussion of candidates for succession except those who were "natural issue of Her Majesty's body" (Ch285)—previously the wording was "legal issue"—leaving open the possibility of a bastard successor (S26). July: Cecil had an engagement party for his daughter Anne and Oxford (S22). Oct.: 2nd Earl of Southampton was arrested again, for the Ridolfi plot to assassinate Elizabeth (A49), and put in the Tower of London with no visitors allowed (S38). Francis Walsingham became English ambassador to Paris (N69).
1572		John went to London on Stratford business (P269). John was charged with illegal wood dealing at Westminster (P269).	*O Compass*, his first known poem, was written. He published a Latin preface to a Latin edition of Castiglione's *Courtier*.	Feb.: Southampton still held in the Tower, with no charges levied (S39). The Duke of Norfolk was executed for treason. Aug: 10,000 French Protestant Huguenots (some estimates are 30,000) were massacred across France over several days, starting on St. Bartholomew's Day.

TABLE 1 (continued)

YEAR	SHAKESPEARE	SHAKSPERE	OXFORD	OTHER
1573		A warrant was issued for John for a 30-pound debt, in which he was named as a whitener of skins (P269).	Rumored to be Queen Elizabeth's lover. Jan.: The Queen and Oxford visited Matthew Parker, the Archbishop of Canterbury, her mother's former advisor (S33). His first printed poem, *From the Earl of Oxenforde to the Reader,* was recorded in a letter to Thomas Bedingfield, who had written a Preface to the Earl in his book *Cardanus Comforte* (a translation of a Cardano essay) (N77). de Vere's men assaulted his father-in-law's servants on the road to Rochester ["Oxford's Men" go on wild escapades similar to "Prince Hal" and on an identical road]. Asked for naval deployment and refused. Lived in "The Savoy," a literary centre in London near Burleigh's house (L)— Gabriel Harvey and John Lyly also lived at the Savoy. Oxford was noted as being in arrears for 2 rooms at the Savoy (L). Oxford's poems *The Hundredth Sundrie Flowres* and *Revenge of Wrong*" contained phrases that showed up in *Richard III* (C264). Uncle Golding enrolled in the Inner Temple London. Lord Hatton, a favorite of the Queen, wrote to her that Oxford was a "boar" (the boar is on Oxford's coat of arms).	Jan.: Pool wrote "The Queen wooed the Earl of Oxford but he would not fall in" (S26). 2nd Earl of Southampton's wife was pregnant (S39). The Queen visited the Archbishop of Canterbury with Oxford (S33). Francis Walsingham became Principal Secretary and spymaster to the Queen. Feb.: 2nd Earl's wife wrote to the Queen asking for conjugal visits, which the Queen refused (S39). 2nd Earl wrote to Privy Council asking for conjugal visits, which it refused (S39). May: 2nd Earl was released and placed with his wife at his father-in-law's estate under supervision of William More (S39). Gilbert Talbot reported to his father that Oxford was the favorite of the Queen and that Lord Burleigh "winketh at all these love matters and will not meddle in any way" (S27, A67). Mary, Queen of Scots, sent a letter to Elizabeth: "Even the Count of Oxford dared not cohabit with his wife for fear of losing the favor which he hoped to receive by becoming your lover" (S27). July: Queen visited Archbishop of Canterbury again (S33). Sept.: Queen spent her 40th birthday with the Archbishop, then dropped from view for 6 months (S33). Oct.: The 2nd Earl wrote to More that his wife had given birth unexpectedly to a boy and apologized for not informing him in time for More's wife to be present as promised (S39). Henry Wriothesley was born.

TABLE 1 (continued)

YEAR	SHAKESPEARE	SHAKSPERE	OXFORD	OTHER
1574	*Famous Victories of Henry the Fifth* was written (C8) and included the character Prince Hal, which historian Captain Bernard Ward believed was Oxford writing himself into the play (C8). The play also includes an event on May 20th (King Henry had no May 20th in his reign) that actually occurred in England when Oxford's men attacked Burleigh's messengers on the road (C8) on May 20. This play also had the real-life 11th Earl of Oxford as the hero who organized the palisade of stakes that won the war (C15), even though the hero's name was not recorded elsewhere (although that Earl of Oxford *was* an adviser to King Henry). This play seems to be an early version of *1 Henry IV* and *2 Henry IV* and *Henry V* (C13).		June: Asked Burleigh for entry into military service and denied, but he went to Flanders anyway to fight Spain (L). The Queen had Burleigh bring him back (L). Mar.: The Queen visited Archbishop Parker again with Oxford (S33). July: Oxford and the Queen fought publicly (S33) and he left for the Continent. July: His friend Thomas Bedinfeld brought him back (S34). Aug.: Francis Walsingham's diary reported that Oxford was back in the Queen's favor. Sept.: Anne Cecil de Vere, 18 years old, wrote to the Lord Chamberlain and asked for an extra room for her husband in Hampton Court: "...is the willinger I hope my Lord my husband will be to come thither..." (S34). Oct.: Anne and Oxford spend a month at Hampton Court, seat of government (S34), and Burleigh's diary recorded that they have an "extra room".	April/May: The Queen was in seclusion for 9 mos and reportedly "melancholic about weighty matters" (S33). Sonnet 33 mentions a baby being taken away by the "region cloud"—Regina Elizabeth (S34). July: 2nd Earl of Southampton placed on the commission of the Peace for Hampshire, made Commissioner for the transport of grain, a Commissioner of musters, and a Commissioner to suppress piracy (S39). July: The Queen fought publicly with Oxford; he left for the Continent without permission, an act of treason (S33). Oxford brought back by friend Bedinfeld and rejoined the Queen on the "summer progress" in Bath (S34). A reunion is shown in Sonnet 154 (S36).
1575	Performances of *All's Well That Ends Well* and *Love's Labour Lost* at Court (C162).	John bought 2 houses, one next door to his house on Henley St. (P269).	15-month continental tour (S34), with the Queen's permission. Declared that if his wife was pregnant it was not by him (S34). Feb.: Wife Anne felt the quickening. Mar.: Attended coronation of French King Henri III at Rheims and Commedia d'elle Arte entertainments (A76). Apr.: Met John Sturmius in Strasbourg—a polymath, university rector, and intellectual leader of Protestantism (N125). Visited Milan, Siena, Venice, Padua, Sicily, Palermo (N131). Lost his first income property. July: Daughter Elizabeth born (S34). Sept.: Oxford heard of the birth while in Italy (S34). Welbeck Abbey (C29) and St. Albans (N124) portraits painted of Oxford.	Mar.: When the Queen hears that Anne Cecil de Vere is pregnant, she appeared pleased (S34). May: Archbishop Parker dies (S34).

TABLE 1 (continued)

YEAR	SHAKESPEARE	SHAKSPERE	OXFORD	OTHER
1576-1579	11 plays were performed in early versions by Lord Chamberlain's Men at Richmond Court (C21–22).		In the company of Bohemian literary men and play actors (L). Published early lyrics (L). Letter to Bedinfeld about his rivalry with Philip Sidney (L). While selling off properties to pay debts he asked friends for help but was denied, as in *Timon of Athens* (C26–40). 1576–1578: Oxford poem about the loss of his good name was similar in content to Shakespeare Sonnets #71, 72, 81, 112, 121 (L). This topic did not appear in other Elizabethan writings.	
1576	*Wrights History* said Lord Essex identified Oxford as Bertram in *All's Well That Ends Well* (L). Jan.: *The History of Error* (*A Comedy of Errors*) appeared (C19). Feb.: *The Historie of the Solitarie Knight* (*Timon of Athens*) (C19); *The Historye of Titus and Gisippus* [mistranscribed] (*Titus Andronicus*) which includes the Nov. 4 "Spanish Fury" massacre in Antwerp and other Catholic conspiracy events (C43–52). The character Lucius said he is reading Ovid's *Metamorphosis* given to him by his mother (C53).	John applied for a coat of arms (P269) and stopped attending Council meetings (P269).	His poem *O Compass* was reprinted. His letters written from Paris contained particulars suggestive of *Othello* (L). Estranged from his wife (L). Apr.: While in Paris hears of wife's rumored infidelity from Henry Howard (similar to Iago) believes it, goes back to England, separates from wife for 5 years (S34, S37). Taken by pirates, stripped naked, robbed, but recognized by a Scottish sailor and not killed (N137). Oxford demanded restitution from the Prince of Orange. Apr.: Queen sent escorts to Dover to welcome Oxford and confiscated his Italian garments. He wore Italian clothes even though he was ridiculed at court. *Stowe's Annales* in 1615 reported that Oxford had a new stylishness of dress after his return from Italy (N229). Apr.: Lived near Theatre Inns and near The Theatre (S47). Wrote a letter to Burleigh that he wanted nothing to do with his wife and had the Queen bar her from Court when he was there (S47). June: Published *The Paradise of Dainty Devices* (8 poems) (L, N157, S47, A31), a best-seller for decades (A121).	2nd Earl of Southampton's home was taken over by Thomas Dymock at the same time that Henry Wriothesley, future 3rd Earl of Southampton (called Southampton throughout this Timeline) entered the household after being wet-nursed (S40).

TABLE 1 (continued)

YEAR	SHAKESPEARE	SHAKSPERE	OXFORD	OTHER
1577	Jan.: The masque *A History of Error* (early version of *A Comedy of Errors*) performed at court for the Queen by St. Paul's Boys (S47). Feb.: *The Historye of Titus and Gisippus* (early version of *Titus Andronicus*) performed at Whitehall by St. Paul's Boys and *The Historie of the Solitarie Knight* (early version of Timon of Athens) performed at Whitehall Palace by St. Paul's Boys (S48). Holinshed published *Chronicles*, on which many Shakespeare history plays are based (S48). Dec.: *Pericles, Prince of Tyre* performed containing many facts of Oxford's life such as his sea voyage and shipwreck (C60–62), his daughter being born while he was away on the Continent, and his estrangement from his wife thereafter (C58). The play ends with the reunion of *Pericles* and his wife and his seeing his 2-year-old daughter (C74).		Jan.: The masque *A History of Error* (early version of *A Comedy of Errors*) performed at court for the Queen by St. Paul's Boys. Oxford leased Hayridge in Devon to Robert Seas for 2,000! years. Dec.: Duchess of Suffolk wrote to Burleigh that she heard Oxford was buying a house in Watling St. London and will not continue as a Courtier (S48). Dec.: Oxford's half-sister married Peregrine Bertie whom the Queen sent as ambassador to Denmark 6 years later. The things Bertie observes there show up in *Hamlet* (S48).	First public space for theatre opened north of London, and was called The Theatre.

TABLE 1 (continued)

YEAR	SHAKESPEARE	SHAKSPERE	OXFORD	OTHER
1578	Jan.: *The Rape of the Second Helen* (early version of *All's Well That Ends Well*) performed at Court. *A Morrall of the Marryage of Munde and Measure* (*The Taming of the Shrew*) (C94) was one of the plays set in Italy (C95), and *A Double Maske—A Maske of Amasones and an Other Maske of Knightes* (*Love's Labour's Lost*) was performed. Mar.: *The History of Murderous Mychaell* (early version of *Arden of Feversham* with Oxford acting in the play (C116)) had some unique Shakespearan terms such as "jets" (a way of walking), which also appeared in 3 other Shakespeare plays but nowhere else (C121). *Arden of Feversham* and *The Famous Victories* both included references to a theft on May 20 on Gad's Hill, the real-life place where Oxford's men attacked messengers of the Court (C17). Dec.: *An History of the Cruelties of a Stepmother* (early version of *Cymbeline*) (where the character Posthumous seems to be the real-life Oxford) and the subject was the marriage negotiations of Queen Elizabeth and duc d'Alençon (C162); performed at Richmond Court by The Lord Chamberlains Men (S49). Alluded to in literature: *Taming of the Shrew*, *Measure for Measure* (about the revival of blue laws defining strict codes of dress for the classes and no excess such as foreign fabrics) (C24).	John mortgaged his wife's house at Wilmecote to relative Edmund Lambert (P269). John was excused the poor tax (P269). John sold part of his property (P269). John was sued for 30 pounds and noted as a whitener of skins (P269). There is no evidence that his son William ever went to school.	Jan.: Letter to Burleigh complained about Burleigh's advice not to sell lands, and said Oxford must sell them since the Queen would not give him a military commission nor any other (paid) position in government (C478). This letter included some of the same words from *Hamlet* about promises and waiting "while the grass grows," and "starve while the grass dost grow" and "I eat the air; promise-crammed, you cannot feed capons so," and "Sir, I lack advancement" (C478). Sept.: Frobisher's third voyage to find gold in the New World returned with worthless ore, and investor Oxford lost his investment of 3,000 pounds (C191, S48). He accused Michael Lok of swindling him (N187), and the Court agreed and sent Lok to Fleet Prison (S49). [Antonio took out a 3,000 ducat bond in *The Merchant of Venice*.] Oxford came into control of another 22 properties from his inheritance. Lyly started working as his Secretary, and both of them lived at the Savoy (L, N183).	The idea of the Virgin Queen was first mentioned in an entertainment by Thomas Churchyard. May: duc d'Alençon wrote to Queen Elizabeth of his affection for her (C163).

TABLE 1 (continued)

YEAR	SHAKESPEARE	SHAKSPERE	OXFORD	OTHER
1578–1579	[The History of] *The Rape of the Second Helen* (early version of *All's Well That Ends Well* and of *Love's Labor's Won*) performed at court (C102, S49) with *Love's Labour's Won* (first performed as a *A Double Masque*) for the envoy of the duc d'Alençon (C168).		Oxford was a founding member and patron of the Euphuist school of poets.	Jan.: The Queen wrote to duc d'Alençon about his possible visit, and entertained his marriage proposal (C162) for several years. The question of the Queen's marriage was addressed in 12 of Shakespeare's plays (C433).
1579	Feb.: *The History of Serpedon* [mistranscribed for *Cleopatra*] (*Antony and Cleopatra*) (C202), and *The History of Portio and Demorantes* (early version of *The Merchant of Venice*) (C191) was shown at Whitehall on Candlemas Day night by the Lord Chamberlain's Men (C191, S49). Mar.: *The History of Murderous Michael* by Oxford (early version of *Arden of Feversham* and early version of *2 Henry VI*) performed at Court by The Chamberlain's Men (S49). Stephen Gossen wrote in "School of Abuse" that he saw *The Jew* (another early version of *Merchant of Venice*) in July and *Ptolome* (early version of *Antony & Cleopatra*) performed at the Bull Theatre (C191, S49). Dec.: *A History of the Duke of Millayn and the Marques of Mantua* (early version of *The Two Gentlemen of Verona*) performed at Court by Chamberlain's Men (C162,S50) on the topic of marriage negotiations between Queen Elizabeth and duc d'Alençon. *The Merchant of Venice* was described in contemporary literature.	John was in financial difficulty and mortgaged his wife's estate Asbies and took in a paying tenant (S85). He also mortgaged other properties (P269).	Quarreled with Philip Sidney at a royal tennis court. *The History of Murderous Michael* by Oxford, which seems to be an early version of *Arden of Fenisham*, in turn an early version of the second *Henry VI*), was performed at Court by Chamberlain's Men. Oxford continued to wear Italian clothes and was ridiculed at Court (the blue laws were in effect and outlawed such clothing). He made a request to reclaim the Stewardship of Waltham, an ancestral right from Thomas Clere in the 13th century. This request, among many others for positions in government, was one he continued until he was successful in 1603 (N425).	Feb.: duc d'Alençon's envoy arrived in London (C169). Mar.: Spanish Ambassador Mendoza received by the Queen (C169). Aug.: French Duke arrived in England and pressed his case for marriage to the Queen (C169). Spenser published *Shepherd's Caj* purportedly containing references to Oxford and Sidney as "Willie and Perigot" (L), which was connected to mentions of Willie in his earlier poems (L). 7 plays performed at Court that were written by anonymous aristocrats.

TABLE 1 (continued)

YEAR	SHAKESPEARE	SHAKSPERE	OXFORD	OTHER
1580	Feb.: *Antony and Cleopatra* performed at Court (C213). Plays alluded to in literature: *Taming of the Shrew, Timon of Athens, Anthony & Cleopatra, King John, Twelfth Night, Much Ado About Nothing, 2 King Henry IV*. Anthony Munday, playwright and theatre manager, said he was the servant of the Earl of Oxford (L), and hinted that not all his (Munday) plays were written by himself alone (L), and that they contained passages that "might have rested in the mind of Shake-speare" (L). Dec.: *3 Henry VI* performed at Court, with Oxford's ancestor the 13th Earl of Oxford appearing as a character supporting King Henry, which was true in history (C234). *Coriolanus* was written about Sir Walter Raleigh who was knighted that year for his ship travels and for bringing riches to the Queen (C286). Dean Church in his *Life of Spenser* said 1580–1590 was the period of flourishing for Shake--speare, but Shakspere of Stratford was too young to be this person at 16 years old (L). 1580–1592 Lyly produced plays containing dialogue and experiments later appearing in and foreshadowing "Shake-speare" plays (L).	John fined 40 pounds for missing a court date, and fined 20 pounds as a pledge for his conduct (P269). John sought "sureties of peace" against his creditors for fear of death (P269).	Jan.: Purchased Fisher's Folly mansion for aspiring poets and playwrights, a kind of writing academy and workshop (S50). Over the fireplace, he placed the coat of arms of Southampton (S50). Had a love affair with court lady Anne Vavasour. "Oxford's Boys" players toured the provinces (L) and included in their repertoire Oxford's plays and plays written in part by Oxford (L). Lyly managed the play tours 1580–1584. 1580–1585, Oxford sold off many properties to pay for the literary academy (S50). The heads of Cambridge University wrote to Burleigh objecting to Oxford's servants "showing their cunning" in certain plays they performed before the Queen (L). Oxford turned in Catholic traitors and comrades Howard, Arundell, and Southwell (C233). Organized Oxford's Men players out of the Earl of Warwick's Men (N239) and they went on the road to Norwich, Coventry, Bristol, and Poole (S50). Converted space in Blackfriars Convent into a public theatre featuring choirboy players (S50). 1580–1590: Oxford's Bohemian Period (L). Connected to theatrical manager Anthony Munday, according to England's *Helicon* in 1600, and to the *Shepherd's Joy* poems by Munday (although quality was considered too high to be Munday's work) (L). Oxford's work represented the new realism in English poetry while Sidney represented the earlier more affected and formal style, according to Dean Church's *Life of Spenser* (1879) (L).	Jan.: 2nd Earl of Southampton banished his wife from his household (S40) to another house he owned, and she wrote to her father that she was ill-treated and that Thomas Dymock was running the household (S40). Apr.: The French Ambassador addressed Queen Elizabeth with a message from Catherine de Medici which proposed a joint effort to prevent Philip of Spain from dominating Portugal (C214). Sept.: Ambassador Mendoza of Spain reported that the Duke of Guise recognized James VI as King of Scotland and that relations were strong between France and Scotland (C214). Arrival of covert Catholic missionaries in England. April: Earthquake in London and throughout England, a rare occurrence (C215). Sept.: Sir Francis Drake's *Pelican* returned to England after 3 years, full of riches (C215). In the Fall, the duc d'Alençon accepted the sovereignty of Flanders (C214). The first edition of Montaigne's *Essays* was published (C397).

TABLE 1 (continued)

YEAR	SHAKESPEARE	SHAKSPERE	OXFORD	OTHER
1581	Apr.: The first performance of *A Midsummer Night's Dream* at Court entertained the French Ambassador (C435). Sept.: *Romeo and Juliet* composed, referencing an earthquake that occurred 11 years earlier in Verona. In reality, a severe earthquake did occur in Verona in 1570, which had more than 2,000 aftershocks over a period of more than 3 months (S60). *Richard III* was written, with 26 references to The Tower. *As You Like It* was written, about the duc d'Alençon's courtship of the Queen (C346).	William prosecuted by Sir Thomas Lucy (who had a theatre troupe) for poaching on his game preserves (C432).	Wrote his *Book of Prophesies* (N219). July: Charles Arundel while under house arrest talked about Oxford's book of pictures prophesizing the date of the Queen's death and a "crowned son of the Queen as her successor, Lord Harry." Dec./Jan.: Confessed his Catholic party activities to the Queen, turned in 3 friends as traitors, and was sent to the Tower briefly (N249, S50). 12 days later he gave the Queen an elaborate gift (N261,S51). Jan. or Apr.: Won a jousting contest (S51). Mar./Apr.: Queen throws him and his lover Vavasour in the Tower (C288) for 2.5 months after the birth of their illegitimate son (S51, S60). Vavasour had at least one other lover and public sympathy was with Oxford (C289). Many duels were fought over her by Oxford and others (some of which were political) (C302). Then she married someone else at Court (C312). June: Oxford was released from The Tower, was under house arrest for the rest of the year (S60), and was exiled from Court for 2 years. Vavasour's uncle killed one of Oxford's men in a street brawl. July: Wrote to Lord Burleigh thanking him for interceding with the Queen to effect his release from the Tower (C267). Earl of Oxford's Men played Coventry every year from 1581 to 1585, while other companies played there less often. One of the other companies was that of Sir Thomas Lucy (C432).	Jan.: 2nd Earl of Southampton was imprisoned again due to new anti-Catholic laws (S40). Apr.: French ambassadors arrived in London to negotiate a marriage between the Queen and Hercule Francis duc d'Alençon, youngest son of French King Henry II (S56). Execution of Catholic missionary Campion. June: Earl released from prison and at 35 yrs old wrote out his will with Thomas Dymock as executor (S40) and beneficiary (S44), with control over his child (S44). Aug.: 2nd Earl was arrested again after being accused by Dymock of having been in contact with Campion (S40). Sept.: 2nd Earl released and returned to Titchfield (S40). Oct.: 2: Earl died at age 35 under the care of Dymock. His will disowned his daughter if she lived with her mother (S40) and left a bequest for education to age 21 for "William, my beggars boye" (S46). Dymock lived on a Southampton estate for 20 more years. (S44). Nov.: Queen Elizabeth said she would marry the French duc d'Alençon (C346) and called him her "little Moor" (C388). Nov.: The Earl's wife said she had not seen her son in almost 2 years (S40). Dec.: Henry Wriothesley, 3rd Earl of Southampton, entered the Cecil household as the 8th and last of the Queen's wards (S40, S62) at age 7, where the 16-yr-old Earl of Essex was a ward. Robert Cecil was 18 years old (S56).

TABLE 1 (continued)

YEAR	SHAKESPEARE	SHAKSPERE	OXFORD	OTHER
1582	Jan.: *A History of Caesar* performed at Court (C392). Some of the sonnets were written during 1582–1589. *Richard II* was written (C329). Feb.: *Ariodante and Geneuora* (early version of *Much Ado About Nothing*) was performed before the Queen (C388) and alluded to a previous relationship between Benedick and Beatrice before the play begins (C391).	Aug.: William conceived daughter Susanna (baptized in May 1583) with Anne Hathaway (P270). Nov.: Marriage license was issued for William Shaxpere and Anne Whateley of Temple Grafton (S85, P270). Then the next day a bond was issued to ensure propriety of marriage of William Shagspere and Anne Hathaway of Shottery (P270). They lived on Henley St. with John (S88). John was a witness in a Chancery suit about his inlaw Ardens' estates (P270).	Dueled with Vavasour's uncle Thomas Knyvet and both were injured, with Oxford getting the worst of it and becoming lame for the rest of his life (S60). One of Oxford's servants was killed during the duel (A178, N284). Thomas Watson dedicated to Oxford a book of his 100 sonnets, for which Oxford wrote an introduction fo each sonnet, commenting on the references used in each. These sonnet introductions would be the only known work of literary criticism by the writer Shakespeare, if de Vere was Shakespeare (A182).	duc d'Alençon marriage proposal collapsed: Elizabeth became celebrated as the Virgin Queen. Mar.: Duke of Orange wounded by a Spanish assassin (C399) (alluded to in *Julius Caesar*). Second edition of Montaigne's *Essays* was published (C399).
1583	*1 Henry IV* written (C490). Hamlet, *Price of Denmark* written, an early version of *Hamlet* (C456), including the Ridolfi Plot of 1569 against the Queen (Norway stands in for Scotland), with references to the plague of 1582 and the comet of 1582 (C463). *Othello, The Moor of Venice* was written, with allusions to the Queen's betrothal to the duc D'Alencon (ruler of The Netherlands) (C388). Feb.: *A History of Ariodante and Guinevere* (early version of *Much Ado about Nothing*) played at Court by Mr. Macalster's Children since Oxford was still banished (his players could not perform) (S61). *Cymbeline* alluded to in literature.	May: Daughter Susanna Shakspere baptized, mother was Anne Hathaway (S85). May: John's tenant took him to court to get out of their lease and the tenant won (S85). William's children remained illiterate their whole lives (Prxiii, 244). His daughters could not sign their names at their marriages (R674).	Leased Blackfriars theatre in London and transferred the lease to his Secretary John Lyly (S68). Buried a legitimate infant son. Pardoned by the Queen and readmitted to court (S67). Traveled to Oxford University with the Court. Fought with his wife's uncle and was wounded. Reconciled with wife Anne Cecil and visited Cecil house where Southampton was being tutored. Formed the largest children's playacting company, Oxford's Boys, by joining the Children of the Chapel and the Children of St. Paul's (S67). Oxford's brother-in-law sent to Denmark on a diplomatic mission for 5 months to give King Frederick II the Order of the Garter (Denmark had started taxing British ships going through Danish waters to Russia) (C457). He met with Danish officials named Rosenkrantz and Guldenstern (A191). Some people, places, and events from that residency showed up in *Hamlet* the following year.	Mother of duc d'Alençon urged her son to come home before he became the "laughingstock of the world" (C400). Jan.: Lord Burleigh reported that Oxford was financially ruined and in adversity and had only 4 servants left (S61). Spymaster and Puritan Francis Walsingham created the Queen's Men from the 12 best actors, to spread Protestantism and to heal radical splits within it (S67).

TABLE 1 (continued)

YEAR	SHAKESPEARE	SHAKSPERE	OXFORD	OTHER
1583– 1585			Oxford Company visited Stratford.	For two seasons, all Court performances were by Oxford's Boys (C629) or Queen's Men (C629, S68), and no other companies were allowed to perform at Court (S68).
1584	Jan.: *Felix and Philomena* produced (C456), with many allusions to *Hamlet* (C456). *The Merry Wives of Windsor* written (C511), with Falstaff based on real-life Captain Dawtrey (C511) who put down the Desmond Rebellion. Mar.: *Sappho and Phao* by Lyly (same subject as *Twelfth Night*), with a Puck-like character, given at Court by Oxford's Boys (S68). Dec.: Queen's Men previewed *A Pastoral of Phyllida & Coren* (early version of *A Midsummer's Night Dream*) for the Queen (S68, C435). Dec.: *The History of Agamemnon & Ulisses* (early version of *Troilus and Cressida*) by Lyly performed by Oxford's Boys (C449, S69). *Comedy of Errors* and *Felix and Philomena* (early version of *Two Gentlemen of Verona*) previewed by Queen's Men for the Queen (C435). *Hamlet* first produced (C660) and includes the "I am that I am" speech in Sonnet 121, which also appeared in a letter from Oxford to Burleigh, accusing him of spying on Oxford's household (C484–486). *The Tempest* first performed (C424), included text from Montaigne's *Essays* (C423). *Julius Caesar* performed before *The Tempest*, and included lines later used in *The Tempest* (C417, 423).	April: Twins Hamnet and Judith (baptized in February 1585) were conceived (P270). Late 1584: Shakspere left for London at the same time Oxford's Boys were playing in Stratford-upon-Avon (L4992, 4995, 8035).	Daughter Bridget born. de Vere's troupe performed *The History of Agamemnon & Ulysses* at Court. According to Looney (L), it was written by de Vere. Argued for a commandership in the Lowlands war and was sent there briefly, but then was recalled by the Queen. Leicester and Sidney were sent to command instead (A205, 206) (replayed in *Othello* and in *Hamlet*). Received control of his final inheritance of all the Earl of Oxford properties. Acquired a sublease to Blackfriars Theatre (R21). 1584–1587: Oxford's Boys established in London where they performed plays written by Oxford (L). Oxford worked in his studio at Fisher's Folly (S68). Attended four-day event at Oxford University where Girodano Bruno spoke about the heretical theory of Copernicus that the earth orbited the sun, that the universe was infinite, and that therefore there could be no heaven or hell. Bruno's celestial tenets show up in *Hamlet* (A195). Also present was Polish Prince Laski. A one-time only play was performed, *Dido*, and some lines from it show up later in *Hamlet* such as "How often does the sad shade of my father appear. . . " (A195). The Queen assigned Oxford to a Parliament Committee that considered petitions for exploration of the New World (A199). He invested in the exploration of a northwest passage through Canada, but lost money (N189).	Assassination of Dutch Protestant leader William of Orange. Oct.: Burleigh and Privy Council created "Bond of Association" to guarantee loyalty by swearing everyone to an oath of loyalty to avenge any threat to the Queen (S68). Robert Cecil at 21 years old traveled to France with a list of precepts from his father Lord Burleigh, as Polonius gives to Laertes in *Hamlet* (S62). Robert Cecil sat Parliament for Westminster (S63). Queen named the new colony "Virginia" after herself, the "Virgin Queen" (A200). Walter Raleigh interceded with the Queen and with letters to Burleigh to keep Oxford in good favor (N290–291).

TABLE 1 (continued)

YEAR	SHAKESPEARE	SHAKSPERE	OXFORD	OTHER
1585	Lyly's play *Endymion* contained some lyrics almost identical to lyrics in the *Merry Wives of Windsor,* published in 1632 (L). Jan.: Queen's Men previewed *Felix and Philomena* (early version of *Two Gentlemen of Verona*) for the Queen (S69).	Feb.: William's twins are baptized as Shakspere (S85,P270) and named Hamnet and Judith after his Sadler neighbors.	Given a military commission and sent to the Lowlands (Netherlands) in charge of 4,000 men (N296). Recalled after 4 months (N299). 1580–1585: Sold 32 of 56 properties (77 originally) to pay for his troupe of actors and writers (C473). June: Wrote to Burleigh about the long-requested and long-awaited funds for putting on plays in support of the Queen (C475).	Earl of Leicester (Dudley) became the new favorite of the Queen. Oct.: Southampton entered St. John's College at age 11.5 years for 4 years (S56,62). Essex at age 20 was released early from his wardship, with help from his stepfather the Earl of Leicester, and then entered Parliament (S62).
1586	*The Phoenix and the Turtle* was written. *The Winter's Tale* was written, and included real events from Walter Raleigh's life during 1582–1586 (C541) such as his short-lived settlement of the Virginia colony, and its rescue by Drake (C525). *Arden of Feversham* was performed between 1586 and 1592 (C120). Performances at Court of *The Famous Victories of Henry V*, early version of *Henry IV* and *Henry V*, and *The Troublesome Reign of King John*, early version of *King John* (S71).	John removed from the Stratford Board of Alderman (S85) for 10 years of non-attendance (P270). John issued with order for debt but has no goods to distrain (P270).	Sat on jury trial of Mary Queen of Scots (L). June: Burleigh wrote to Francis Walsingham asking him to speak to the Queen about Oxford's finances—his daughter (Oxford's wife) being worried. June: Queen Elizabeth signed Privy Seal Warrant Dorman, authorizing for de Vere to receive 1,000 pounds a year for life (it ran 18 years until de Vere died), retroactive to March (N301), and taken from Walsingham's spy budget (S69). This annuity continued even after the Queen's death. No one else received such a large outright sum from the Queen.	Queen Mary arrested for the Babbington Plot to assassinate Elizabeth (L). Queen Mary sentenced to death for treason (L). Diplomatic relations cut off with Spain (paralleled in *Henry V*) (C24). Philip Sidney died (L). Robert Cecil sat Parliament for Westminster (S63).
1587	*Two Gentlemen of Verona* written (L). Shakespeare known to be in London. *1 Henry VI* written, updating the Henry VI story with parallels to Mary Queen of Scots [Joan of Arc]. *2 Henry IV* written (about the recent campaign in the Low Countries and the Babington Plot) (C525). *Henry V* (including the Babington Plot) performed by the "University Wits," writers and actors directed by Lord Oxford (C587). Plays alluded to in literature: *1 Henry VI, Richard III, Julius Caesar, Merry Wives of Windsor.*	While his father was being prosecuted, William took care of his mother (R16). Legal action with his siblings against his mother's estate. The only known letter written to Shakspere was in this year, asking for a loan of 30 pounds, which went unanswered (L), or perhaps was never sent.	Oxford's Players (aka Oxford's Men), one of the 4 leading theatre companies in London, was active for the next 15 years (N391). Oxford's dramatic writing activity appeared to stop in this year (L). Jan.: Francis Walsingham's spy reported that Oxford's company was one that put up playbills in the city every day of the week (C629). Daughter Susan was born.	The Queen made the Earl of Essex Master of the Horse and her new favorite, as Leicester was ailing (63). Queen Mary was executed (L). Philip Sidney died. A large, expensive funeral was held for Sidney by his father-in-law Francis Walsingham (L).

TABLE 1 (continued)

YEAR	SHAKESPEARE	SHAKSPERE	OXFORD	OTHER
1588	*Love's Labour Lost* was written. Plays alluded to in contemporary literature: *Troilus and Cressida, Richard II, Richard III, King Lear, King John* twice, *Hamlet, Romeo and Juliet, The Merchant of Venice, Titus Andronicus. Hamlet* was written in 1588–1589.	John and Mary (William's parents) began an unsuccessful suit against John Lambert for return of Mary's property; William was added as a plaintiff (P270).	Participated in the early intercept force against the Spanish Armada (L,N313). June: Wife Anne died (a rumored suicide) (L,S63) at age 31 (L) in Greenwich Palace and was buried in Westminster Abbey (S63). Oxford and the Queen did not attend her funeral (S63). Oxford retired into private life (L) and sold Fisher's Folly.	Mar.: Leicester, despite his poor health, was put in charge of the land army in preparation for the Spanish invasion (S63). Apr.: Essex made Knight of the Garter (S63). Spanish Armada launched from Lisbon. Oct.: English naval forces defeated the Spanish Armada (S63). Oxford was injured in his ship during the battle (S63). War dragged on through 1603 (S63). Puritan, anti-Anglican pamphlets circulated (A240). Earl of Leicester ("Robin") died and the Queen was devastated by the death of this friend since childhood (S63). Mar.: Southampton was admitted to Gray's Inn for law studies (S56). Duke of Guise was assassinated by French King Henry III, which was paralleled in *Macbeth* (C24).
1589	Some Shakespeare works dated to be before 1589 by Hotson and by Brown in 1949 and by Alexander in 1950. Shakspere too young at 13 to be the author of those Shakespeare works (L). Thomas Nashe mentioned the *Hamlet* play. *The Taming of the Shrew* written. Plays alluded to in literature: *Hamlet, The Merchant of Venice, Romeo and Juliet, Othello, 2 Henry VI, 3 Henry VI, Troilus and Cressida, Julius Caesar, 1 Henry I, Merry Wives of Windsor.* Spenser wrote of Willie "from whose pen large streams of honey and sweet nectar flow" which harks back to his 1579 poem about Willie (L).		*Arte of English Poesie* listed de Vere as a court author whose works would be widely lauded if his "doings could be found out and made public with the rest". Writer Puttenham classed Oxford with Richard Edwards as "deserving highest praise for comedies and interludes" (L).	Murder of French King Henri III. Henri of Navarre became King Henri IV. Apr.: Essex went with Sir Francis Drake raiding the coasts of Spain and Portugal without the Queen's knowledge or permission (S63–64). June: Southampton received Master's degree at Cambridge University (S56). July: Essex made overtures to James of Scotland as a natural successor to the Queen (S64). James VI of Scotland married Anne of Denmark ensuring he would remain a Protestant (C600). Robert Cecil sat Parliament for Hertfordshire (S63).

TABLE 1 (continued)

YEAR	SHAKESPEARE	SHAKSPERE	OXFORD	OTHER
1590	1590–1600 English literature changed dramatically because of Shakespeare's plays and the use of the Italian sonnet form. *1 Henry VI* and *2 Henry VI* were written. Orthodox date for the writing of the first Sonnets (L); other authors said they were written between 1582 and 1589. *King John* written before 1590–1591. Plays alluded to in literature: *Hamlet, Titus Andronicus.*	Stratford town was in serious financial distress, and the bailiff and burgesses of Stratford petitioned Lord Burleigh for relief (S86). John's only asset was his house on Henley St. (S86). This year was the beginning of William Shakspere's so-called theatre career as a sometime actor (L).	Spenser wrote *Teares of the Muses* with probable reference to Oxford as "Willie" "sitting in an idle cell" (L) and as "our pleasant Willie who is dead of late" (L) referring to the lack of court performances after Oxford had put on many plays in the 1570s and 1580s. His group of writers were now scattering because of his lack of financial resources. Oxford wrote 17 sonnets to Southampton for his 17th birthday, encouraging him to have a son (S74). A proposed marriage between Oxford's daughter Elizabeth (Cecil's granddaughter) and Southampton was promoted by Lord Burleigh (Cecil) and de Vere (L). Burleigh told Southampton he was a prince (son of the Queen). Promotion of this proposed marriage went on for 3 years.	Robert Cecil became Secretary of State. Principal Secretary Francis Walsingham died. His granddaughter wrote a poem with a hidden acrostic in his memory (R630). Nov.: Elizabeth's Ascension Day (S90) celebrated the "Virgin Queen" idea, which deified the Queen as Gloriana (S90). Oct.: On Southampton's 17th birthday, Burleigh gave him a year to make up his mind about marrying his granddaughter, Oxford's daughter (S74). Spenser published *Faerie Queen* with dedications, including one to Oxford, and then received a . He received a poem from an anonymous "Ignoto" about giving praise where it is due, with a hidden cipher of E Vere (R667).
1591	Aug.: *Love's Labours Lost* was performed by Oxford's Boys at Southampton's estate (S74). *A Comedy of Errors* contained the line "I buy a thousand a year! I buy a rope". *3 Henry VI* and *Richard III* written. Plays alluded to in literature: *Titus Andronicus, King John, Timon of Athens. The Troublesome Raigne of King John* by Gerard Peele or Christopher Marlowe published and performed (C315), and used by Shakespeare as a basis for *Life and Death of King John* (C315).		Made over Castle Hedingham in trust to his 3 daughters. Married wealthy lady-in-waiting Elizabeth Trentham (L). By this time he had lost or sold all 77 properties he inherited (he had been the richest Earl in England).	Robert Cecil was appointed to a seat on the Privy Council (S64). Aug.: The Queen spent 5 days at Southampton's estate. *Love's Labour Lost* was performed for the first time, at a small park on Southampton's estate (S74).
1592–1601	Oxford's earlier plays start appearing as attributed to William Shakespeare, and Lyly's plays ceased appearing (L).	1592–1596 William resident in London (L), and his name appeared as an actor on one playbill (L). He moved several times, possibly to avoid the tax collector.	Retirement from public life. The "Great Blank" in Oxford's record started in this year. Oxford's earlier plays started appearing as attributed to Shakespeare (L).	

TABLE 1 (continued)

YEAR	SHAKESPEARE	SHAKSPERE	OXFORD	OTHER
1592	First theatrical allusion to Shakespeare (L): *Groatsworth of Wit* by Robert Green warns 3 playwrights (Marlowe, Nashe, Peele) of an actor called Shake-Scene. *Two Gentlemen of Verona* written (L). First date for *Love's Labour Lost* (L). *Arden of Feversham* first printed (C121). Plays alluded to in literature: *Hamlet, Titus Andronicus, King John, Romeo and Juliet, Love's Labour Lost, 1 Henry IV, 2 Henry IV, Twelfth Night, As You Like It, Much Ado About Nothing.* Mere's account of Elizabethan literature lists authors and titles of plays, but no titles for Oxford (L).	Mar.: John recorded as absent from Church (P270). John fined for not attending church (S86). Aug.: John hired to appraise his dead friend Henry Field's belongings, Henry's son was Richard Field the London printer (S86, P270). Sept.: John recorded again for absence from church but not fined (P270).	Purportedly, going by the name "Will Monox," Oxford joined Robert Greene and satirist Thomas Nashe on Greene's fateful and fatal day of drinking and overindulgence (R671). Mere's account of Elizabethan poetry listed authors (including William Shakespeare and Edward de Vere) and titles of plays, except no titles were given for Edward de Vere (L). Oxford's plays started appearing as attributed to William Shakespeare (L), and Lyly's plays also ceased to appear under his name. Oxford is named a prominent playwright, but no titles are given in Meres account of Elizabethan playwrights.	Playwright Robert Greene died, apparently of overindulgence. Posthumous pamphlet by Greene apparently lambasted actor Will Shakespeare as a great literary pretender. Historian Meres' account of Elizabethan poetry in this year listed authors and titles of plays including Shakespeare and de Vere. Both Lyly and Munday had work attributed to them which was not theirs (L).
1593	Apr.: First use of the name "William Shakespeare," on the published poem *Venus and Adonis*, which was based on *Ovid's Metamorphoses Book 10*, which he called "the first heir of my invention" (L). Language analysis says it could not be by a Warwickshire man (R632). The poem was stamped by the Archbishop of Canterbury (S79) and printed by Richard Field (S79). The poems were dedicated to the Earl of Southampton (L,S79). July: *Venus and Adonis* reprinted after the first edition of 1,250 copies sold out. Only 50,000 of 200,000 Londoners were literate (S80). *King Lear* and *Twelfth Night* written. Plays alluded to: *Hamlet, Julius Caesar, The Tempest, Romeo and Juliet, Love's Labour Lost, The Winter's Tale, A Midsummer Night's Dream.*		Son Henry born (L). Before his son and heir was born, Oxford had been pushing for Southampton to marry (which is the subject of Sonnets #1–17 dedicated to Southampton, and the subject of the letters from Oxford to Southampton). The letters stopped when Southampton finally refused to marry Oxford's daughter and paid a large fine to Lord Burleigh (the grandfather of Oxford's daughter). Sonnet 2 said: "When 40 winters shall besiege thy brow . . ". This was when Oxford was 43 and Shakspere was 29. Nashe's pamphlet *Strange News* was dedicated to de Vere as "Gentle Mr. William" and used a Cardano Grille to encrypt the name de Vere (R672) and a message "Lo, so test E Ver" (R672).	Essex was appointed to a seat on the Privy Council (S64).

TABLE 1 (continued)

YEAR	SHAKESPEARE	SHAKSPERE	OXFORD	OTHER
1594	*Taming of the Shrew* was first performed (A130). *Rape of Lucrece* poem dedicated to Southampton (L,S90). Plays alluded to: *Hamlet, 1 Henry IV, 2 Henry IV* (early version), *The Winter's Tale.* Shakespeare mentioned for the second time in literature, in *Willobie and His Avisa,* a pamphlet of 72 cantos, in an introductory poem.	Sept.: Stratford town was devastated by fire and most of Henley St. destroyed (S86). Shaksper played the role of Ghost in *Hamlet* (A276). Shakespeare biographer Halliwell-Phillipps in the 1800s checked previous records in 70 towns and cities in England for evidence of Shakespeare to no avail (L).	The pamphlet *Willobie His Avisa* was published, with Avisa purportedly representing Oxford's wife Elizabeth (suggesting a scandalous affair between her and Southampton—H.W.—with de Vere as W.S. satirically portrayed as egging on Southampton).	The last line of Sonnet 94 also appeared in *Edward III* before 1594. Nov.: Southampton paid Lord Burleigh (Cecil) 5,000 pounds as a penalty for not marrying his granddaughter (S90, A279). Southampton's widowed mother married into the Cecil clan (Thomas Heneage).
1595	Jan.: 3rd edition of *Venus and Adonis* published (S90). *Richard II* performed by Chamberlain's Men, and included the handing over of the throne to Bolingbroke (S91). Mar.: First historical mention of Shakespeare in connection with theatre: with Kempe and Burbage, as a payee of the recently formed Chamberlain's Men for performances at court in Dec. (S96). Literary mentions of *The True Tragedy of Richard, Duke of York, Death of King Henry VI* (early version of *3 Henry VI).*	Mar.: William received payment, with Richard Burbage and William Kempe, for a play for the Queen at Greenwich at Christmas (Pr31,P270). Sept.: Stratford town again devastated by fire, and that time the Henley St. house was completely destroyed (S86) and John seems to have lost everything (S86). Dec.: Stratford town again petitions the Crown for relief, as one-third of the population were paupers, there was a soaring death rate, and vagrants were denied entry to the town (S86).	Daughter Elizabeth married the 6th Earl of Derby, William Stanley (N349). In celebration of the marriage, *A Midsummer Night's Dream* was presented at Court to the Queen. It had been performed at Court previously, in 1581 and 1584 (C435). Thomas Edwardes' *Narcissus* alluded to de Vere as being Shakespeare (A181).	Southampton (21 yrs old) became a favorite of the Queen (S91).
1596	Thomas Lodge referenced the *Hamlet* play as being 4 hours long. *The Tempest* was written. Plays alluded to in literature: *Hamlet, Macbeth, Othello. Romeo and Juliet* first published, anonymously (C314), and the play has many phrases similar to Oxford's poems (C311-314). *Romeo and Juliet* did not have the name Shakespeare associated with it until the 1623 First Folio (C314).	Aug.: Son Hamnet buried. (P270). Recorded as living in Bishopsgate, London (P270). Pursued for 5 shillings for London taxes. Reapplied with his father for a coat of arms, which had been rejected 28 years before (S96, P270)—the fee was 30 pounds (S96). A coat of arms was granted but later rescinded. 3 plaintiffs applied for protection from him (Pr36). Pursued for 5 s for London taxes (P270). Nov.: Legal writ in Southwark against him and 3 others, to keep the peace (P270).	Oxford's wife purchased King's Place in Hackney. de Vere, his son, and his wife moved in.	Earl of Essex led a successful raid of a Spanish outpost in the Azores. July: Robert Cecil was made principal secretary to Queen Elizabeth (S96) and Secretary of State (S64) and he also ran the Secret Service. Earl of Essex led a failed raid of the Spanish fleet at Cadiz. Oct.: Francis Bacon warned Essex that Robert Cecil was plotting his downfall (S96).

TABLE 1 (continued)

YEAR	SHAKESPEARE	SHAKSPERE	OXFORD	OTHER
1597	Jan.: *Richard II* was first published (S96), anonymously. Plays alluded to in literature: *Hamlet, King Lear*. Booksellers tried to secure the copyright of plays (L).	May: Purchased New Place, Stratford, for 60 pounds, second largest house in Stratford (L,P270). Nov.: Reported in London for default of 5 s taxes (P270). Bequeathed his sword to friend John Combe's nephew. John restarted the case over the lost Wilmecote lands (P270).		July: *Isle of Dogs* by Jonson and Nashe was performed at the Swan Theater, and then suppressed (because of an allusion to an island in the Thames where the Privy Council met) (S97); 3 players were arrested: Spenser, Shaw, and Jonson (S97), and Cecil closed all theatres in London (S97). Ben Jonson went to prison for killing Gabriel Spenser (S97); he escaped execution by claiming the "benefit of clergy" by proving he could read Latin (S97); he converted to Catholicism, was convicted, branded on his thumb, and released.
1597-1604	Great period of Shakespearean publication (L).		1592–1601: Oxford's retreat from public life, his "Great Blank". Poems and plays stopped being attributed to him after 1593 (N385).	
1598	Shakespeare's name was first printed on plays (L,Pr144, A259). 17 plays were written by this year. Ben Jonson dated *Titus Andronicus* to this year. Sept.: Historian Francis Meres in *Palladis Tamia* said Shakespeare was the author of 12 plays previously published anonymously (S98). Oct.: *Richard II* and *Richard III* republished with Shakespeare's name (S98). Joseph Hall in *Biting Satires* said the author of the poems was "Labeo," and John Marston wrote the same thing in *Pgymalion's Image*. Labeo in Roman times was the fake name of a writer to hide the true author's aristocratic pedigree. The Archbishop of Canterbury recalled the books and had the Marston book burned.	Reported living in St. Saviour's Parish London. Legal citation as a tax defaulter (L). Jan.: Richard Quiney asked him about investing in Stratford land. Feb.: Recorded owning corn and malt at a time of shortage (S97). Lived at New Place Stratford (S97). Received 10 d for a load of stone at Stratford. Received 20 d for wine to host a visiting preacher. Oct.: Recorded as a defaulter on London taxes. Oct.: R. Quiney letter (not sent) to Shakspere asking him for a 30 pound loan, the only known letter addressed to Shakspere (L). Nov.: A. Sturley wrote to R. Quine urging pursuit of a loan from Shakspere. Noted as hoarding 80 bushels of malt at a lean time.	Feb.: Oxford was presented to the French King by Robert Cecil on an official visit to the French Court. Sept.: Francis Meres (whose brother-in-law was a tutor of Southampton) named Oxford as #1 of 17 comedy playwrights, in his *Palladis Tamia* (*Wit's Treasury*), a catalog of contemporary writing and art. He said "The best for comedy among us be Edward Earl of Oxford" (C638). Shakespeare also was mentioned in this *Who's Who* volume and compared with Ovid (S100). Oxford's son-in-law the Earl of Derby was reported to be writing comedies.	Aug.: Death of William Cecil, Lord Burleigh at 77 years old (N371,S98); his son Robert took over as advisor to the Queen (S98). Historian Ward commented that Burleigh had "a career as a Minister to the Crown which has never been equalled in English history. . . Lord Burleigh's unfailing kindness to Oxford. . . Lord Oxford was hopeless as a family man. . . The ruling passion of his life was poetry, literature, and the drama" (C665). Robert Cecil officially presented Southampton to the French King in France (S98). Nov.: Southampton imprisoned on his return to England after marrying illegally (A310).

TABLE 1 (continued)

YEAR	SHAKESPEARE	SHAKSPERE	OXFORD	OTHER
1599	Apr.: *Henry V* performed at Curtain Theatre—a temporary venue while the Globe was being built with wood from the torn-down "The Theatre" (Pr98, S97). *Othello* alluded to in literature. May: The Globe Theatre was completed (S98) and owned 1/10 each by Wm. Shakespeare, Augustine Phillips, Heminges, Page, Kempe, and 50% by the Burbage Brothers. Some sources added in Nicholas Brend as a Globe owner. Sonnets 138 & 144 published in *Passionate Pilgrim*, which said the author's best days were past (Oxford was 49; Shakspere 34). Author dwelled on his aging in Sonnets 62 and 63, and impending death in Sonnets 66, 71–74, 81. Complaints of lameness in Sonnets 37, 66, 89. Sept.: First performance of *Julius Caesar* at The Globe (S104). *Twelfth Night* appeared in a songbook. Sonnet 63: "when my glass shows me myself indeed,/ Beated and chopp'd with tann'd antiquity" (Pr274).	Recorded as owing taxes in Billingsgate (P271). Feb.: Willelmum Shakespeare with others became a shareholder in The Globe (P271). Refused the right to join his arms with the Park Hall Arden arms (P271). Recorded as owing taxes in St. Helen's Parish, London (P271). Recorded as owing taxes in Clink in Southwark (P271). Recorded hoarding corn and malt at a lean time (P271).	Oxford's son-in-law the Earl of Derby (Darby) was reported to be writing comedies professionally (N393). Apr.: Oxford wrote a speech praising Essex which was inserted into *Henry V*, currently being performed at the Curtain Theatre (S98).	Jan.: Ben Jonson's *Every Man Out of His Humor* contained the character Sogilardo who was ridiculed for getting a coat of arms of a boar without its head [The Oxford crest is a boar] (S102–103). May: Earl of Essex sent to Ireland to defeat Tyrone and failed (S104). Essex brought Southampton to Ireland as the Master of the Horse, against the Queen's wishes (S104). Essex attempted a truce with Tyrone of Ireland (not agreed to by the Queen) and was arrested on his return to England. Aug.: Fears of Spanish invasion, chains drawn across London streets. Queen was dangerously ill (S104). Sept.: First performance of *Julius Caesar* at The Globe (about conspiracy and civil war). A Jesuit spy reported that the Earl of Derby was busy writing comedies. Nov.: Privy Council proclaimed official denunciation of Essex (S104). Dec.: Essex took ill and the Queen sent 6 of her physicians (S105).
1600	*Macbeth* alluded to in literature. John Davies of Hereford wrote an epigram that called William Shakespeare "our English Terence": Terence was a Roman slave used to cover the identity of aristocratic writers such as Scipio and Laelius. He also said Shakespeare did not get his proper honor. 6 plays by Shakespeare were published (L).	Recorded as hoarding corn and malt at a lean time. Willelmus Shackspere sued John Clayton in Queen's Bench for 1592 loan of 7 pounds (P271). Oct 6: Charged tax arrears of 1 mark in London (P271).	Sought the Governorship of the Isle of Jersey again (N394), to no avail.	Aug.: Essex set free but never again allowed in Court (S105) and under house arrest at home with Robert Berkeley. His family was not allowed to live with him (S105). Dec.: Essex and Southampton sent secret letter to James about Cecil (S105). Essex was stripped of offices and placed under house arrest.

TABLE 1 (continued)

YEAR	SHAKESPEARE	SHAKSPERE	OXFORD	OTHER
1601	Feb.: Special performance of *Richard II* at the Globe Theatre paid for by 3 Essex supporters (S105), performed by the Chamberlain's Men with an added scene showing the passing of the crown to Bolingbroke—which had previously happened off-stage (S106–107). Plays alluded to in literature: *Pericles, Othello, The Tempest). Troilus and Cressida* performed. *Twelfth Night* performed in Middle Temple of the Inns of Court. Aug.: *As You Like It* entered into the Stationer's Register (C365). "Our fellow William Shakespeare" lampooned in a Cambridge University play.	2 legal documents named Richard Burbadge and William Shackspeare gent as occupying the Globe. Mar.: Thomas Wittington's will bequeathed to the poor the 40 shillings he was owed by Shakspere's wife (P271). Renewed his father's application for a coat of arms (P271), and received the coat of arms from William Camden (author of *Britannica* and *Remains of a Greater Work Concerning Britain*, which works did not mention him). Sept.: Father died as Shakspere with no coat of arms (P271). In his monument (as a former Chief Bailiff he was eligible for one), his effigy was holding a woolsack, and this monument would later be re-used for his son William.	Oxford emerged from "retirement" to take part in the trials of Essex and Southampton (L). He wrote to Cecil of his poor health and the weakness of his lame hand making it hard to write, although his handwriting appeared to be clear and confident in the letter (N401). He wrote to Cecil seeking support in his bid for the Presidency of Wales (N396).	Feb.: Essex and Southampton rebelled against Elizabeth (and Cecil) and lost (L,C669,S105). Feb.: Jury headed by Oxford condemned Essex and Southampton for treason (S109), and Essex was beheaded on Feb. 25 (S109). Mar. 19: Southampton's life was spared (S110) with no recorded explanation (S109), but he remained in the Tower (S109). Shakespeare Sonnets written to Southampton, possbily while he was in prison (S110).
1602	Date assigned to *Hamlet* (L). *Merry Wives of Windsor* printed. Pirated edition of *Merry Wives of Windsor* published (L). *All's Well That Ends Well* performed.	Complaints made against the Herald (William Camden) for misapproving 23 coats of arms, including the one for John Shagspere of Stratford. Purchased 107 acres and bought a cottage. Manningham recorded joke about William and Burbage as actors (P271). Named a "player" in draft coat of arms (P271). May: Bought land in Stratford for 320 pounds, with brother Gilbert standing in at contract signing (P271). Legal proceedings over New Place in Stratford deeds (P271). Thomas and Lettice Greene took an apartment in New Place (P271). Sept.: Bought cottage and land in Stratford for 80 pounds (P271).	His moribund troupe of actors merged with the Earl of Worcester's Men, who were listed as performing at Boar's Head Tavern (L). Oxford's servants also played at the Boar's Head Tavern.	Southampton was still in the Tower of London prison (L). There is a blank in the accounts of the "Treasurer of the Chamber" (L) for the Tower for that time period.

TABLE 1 (continued)

YEAR	SHAKESPEARE	SHAKSPERE	OXFORD	OTHER
1603	*Hamlet* unauthentically published (L). *Hamlet* printed. John Sanders portrait of Shakespeare was painted (looks nothing like the Shakespeare busts and drawings). Last of the Shakespeare Sonnets were written (L). Isle of Wight referenced in Sonnet 106 (L). *King Lear* alluded to in the contemporary literature. *Henry VIII* written in an un-Shakespearean style, 13 years after *King Lear* (C623).	Listed in papers creating the King's Men troupe of actors. Employed as a marriage broker. Named by James I as Groom of the Chamber (P271). Mar.: Named as a member of the newly formed "King's Men" (P271).	King James renewed the 1,000 pound annuity for Oxford.	Southampton arranged a performance of *Love's Labour Lost* for the Queen (L). Mar.: Death of Queen Elizabeth (L,S112). There was no tribute from Shakespeare or Oxford. Oxford wrote a private condolence letter to Burleigh. The Accession of King James VI of Scotland. Coronation of King James VI, where Oxford performed a ceremonial role (L). Apr.: James's first act as King was to liberate Southampton from the Tower (L,S112,A346). King James gave him an official pardon in May, and wrote in a letter that "the Queen was moved to exempt [him] from the stroke of justice". Apr.: The Queen's funeral was held, and the Tudor reign ended (S112). Southampton was considered for the Knight of the Garter but was instead made a Captain of the Isle of Wight (S112). July: Southampton was made a Knight of the Garter (S112). July: Southampton was made an Earl again and his properties were restored (S112). Cecil received a pension from the Spanish government sometime during James' reign. Ben Jonson started writing masques for King James's court.

TABLE 1 (continued)

YEAR	SHAKESPEARE	SHAKSPERE	OXFORD	OTHER
1604	*Measure for Measure* was first performed (L). The long, official version of *Hamlet* was published officially and some believed it to be Oxford's autobiography (S113). 1604 date was assigned to *Othello* (L). Last of authentic Shakespeare works to be published for 18 years (L). Southampton connection to Shakespeare ceased (L). Nov. thru Feb.: 8 Shakespeare plays performed at Court (C658).	Mentioned as one of the King's Men actors. Sold malt in March–June. Loaned 2 shillings to Phillip Rogers. Retired to Stratford. Rented lodgings from the Mountjoys in Cripplegate. Was issued his "red cloth" for a royal procession of James I into London (P271). Sold malt to Phillip Rogers. He sued to recover the loan from Rogers plus damages of 1 pound 15 s (Pr18). A neighborhood survey recorded his growing real estate empire. He took legal action (L) to force payment for malt he had been supplying. Lodged with the Mountjoys in Silver Street London and negotiated a marriage settlement for their daughter (P271). Oct.: Leased a cottage at Rowington London for 2 s 6 d per week (P271). July: Sued Mr. Rogers of Stratford for debt of 35 s for 20 bushels of malt.	June: Edward de Vere died at King's Place (L) of plague. No memorial, no will. His widow took out no Letters of Administration (N194,431), perhaps because there were no assets and only debts. All 77 properties he had inherited were gone (N191). His son Henry became the 18th Earl of Oxford.	King James procession through London, where Southampton was prominently displayed with his mother (S113). June: After Oxford's death, Southampton was arrested and thrown into the Tower and his papers were searched (S113).
1605	William Camden's book about English history, culture, and language, *Remains of a Greater Work Concerning Britain*, in the chapter "Poems" listed 11 modern English poets "whom succeeding ages may justly admire", including Shakespeare.	July: Invested 440 pounds in interest-bearing tithes in corn, hay, wool, and grain tithes in Stratford (P271,S118), that entitled him to burial in the church chancel. The actor Augustine Phillipps bequeathed him a 30 shilling gold coin (P271), the same amount went to Condell, and larger amounts to Heminges and Burbage (P271).	Daughter Susan married the Earl of Montgomery Philip Herbert (N429) and performed in Jonson's *Masque of Blackness* at Court (S117). Later the First Folio was dedicated to Herbert and Montgomery.	Gunpowder Plot to overthrow King James and replace him with his daughter Elizabeth who was 9 yrs old, was foiled by Cecil and Jonson.
1605–1608	Suspension of Shakespearean publications (L).			
1606	*The Two Noble Kinsman* alluded to in the contemporary literature.	Jan. 21: Shown owing Mr. Hubaud of Stratford 20 pounds (P271).		

TABLE 1 (continued)

YEAR	SHAKESPEARE	SHAKSPERE	OXFORD	OTHER
1607	William Camden's *Britannica* in Latin described English counties and towns and their notable inhabitants, with no mention that Stratford was Shakespeare's hometown (Ch129), but he did mention that Philip Sidney had a home there. In Camden's diary, he did not note Shakspere's death, although he did note Richard Burbages' and poet-playwright Samuel Daniel's deaths.	June: daughter Susanna married Puritan Dr. Jon Hall as Shaxspere (P272), and her father gave her a dowry of land. From 1607 on Dr. Hall made personal notes in his treatment records: He described Michael Drayton as "an excellent poet" and said Thomas Holyoak compiled a Latin-English dictionary, and that local schoolmaster John Deep was remarkably pious and learned (Pr236)— nothing about Shakspere (Ch131).	Natural son Henry de Vere was knighted (in this year or in 1610).	Southampton led a parliamentary group to defeat the King's plans for union with Scotland (S118).
1608	Quarto edition of *King Lear*. First time Shakespeare's name appeared on a title page.	Jan.: The Greene's son was baptized and named after him (P272). Aug.: Took a 21-year lease on Blackfriars Theatre (P272), with the Burbage brothers, Heminges, Condell, and Coates. Sued Mr. Addenbrooke of Stratford for 6-pound debt (P272). Sept.: Stood Godfather to William Walker of Stratford.		Robert Cecil became Lord Treasurer for England (S118).
1608–1609	Slight revival of inauthentically published works: *King Lear, Pericles, Troilus and Cressida, Sonnets* (L).			
1609	Sonnets published for the first time in numbered order (S118). Sonnets Dedication said "… eternity promised by our ever-living poet…", "ever-living" means dead—Oxford is dead at this time, and Shakspere is alive.	Pursued Addenbrooke's surety, Mr. Horneby, for 6 pounds (P272). Apr.: Made payment to poor relief in Southwark (P272). Thomas Greene lived in Shakspere's house for a few months. He mentioned his cousin Shakespeare in his diary but not in the context of literature or theatre (Ch130). Greene was a published poet and contributed a "Shakes-pearean sonnet" to Michael Drayton's *The Barons' Wars* (1603).	His widow was given permission to sell King's Place, Hackney.	

TABLE 1 (continued)

YEAR	SHAKESPEARE	SHAKSPERE	OXFORD	OTHER
1610		Legal proceedings confirming ownership of New Place (P272). Completed purchase of 20 acres in Stratford started in 1602 (P272). Legal proceedings over his tithe holdings (P272).		Ben Jonson stopped exhibiting his Catholicism and became a Protestant again (S118).
1611	Nov.: *The Winter's Night's Tale* first produced at Court according to the *Court Revels* (C541). John Davies' pamphlet described Shakespeare as our English Terence (Axxxi). Terence was an actor who served as a front man for hidden aristocratic playwrights in Roman times.	Contributed to cost of Stratford Parliamentary Bill (P272). Leased Stratford barn to Robert Johnson for 22 pounds (P272). Issued bill to recover the Combe family's default on rent (P272). Interest from his local tithes income was 60 pounds (P272). May: Greene left New Place.		
1612	First production of *Macbeth* at the Globe (A400). Inauthentic publication of 3 plays and the Sonnets. Michael Drayton wrote a book including histories of English counties—Drayton was a patient of Dr. Joseph Hall (Shakspere's son-in-law)—but did not mention Shakespeare as a Warwickshire man, only as a "good comedian" (Ch130). Henry Peacham's book *Minerva Britanna* implied a hidden writer for Shakespeare.	May: Witness in Belott-Mountjoy case (P272), name on testimony was Willm Shakp (P272). Completely retired from London to Stratford (L). Feb.: Brother Gilbert buried as Shakspere (P272).	Wife Elizabeth Trentham died (L).	Robert Cecil died (S118). Henry Stuart, Prince of Wales, died (S118), leaving the unpopular Prince Charles in line for the throne.
1613	June: Globe Theatre burned down during the first performance of *Henry VIII* (A401). All the play manuscripts therein were destroyed (C676).	Jan.: John Combe of Stratford left him 5 pounds. Feb.: Brother Richard buried. Mar.: Bought Blackfriars Gatehouse for 140 pounds (P272). Mar.: Took 60-pound mortgage on Blackfriars Gatehouse (P272). Received 44 s (as did Burbage) for impresa for 6th Earl of Rutland. June: Globe burned down. Oct.: Took a share of the lease on the Globe's new site (P272).		

TABLE 1 (continued)

YEAR	SHAKESPEARE	SHAKSPERE	OXFORD	OTHER
1614		Apr.: Made payment for poor relief in Southwark. Thomas Greene lived in Shakspere's house and named his children after William and his wife. He mentioned his cousin Shakespeare in a letter. Sept.: Noted as owning 127 acres of land in Stratford (P272). Oct.: Given surety against losing tithe income (P272). Nov.: In London with son-in-law John Hall to meet Stratford Town Clerk Thomas Greene over enclosures (P272).		Ben Jonson wrote the masque *The Golden Age Restored* (Elizabeth's Age) (S118).
1615	Susan de Vere Herbert's brother-in-law the Earl of Pembroke won appointment as Lord Chamberlain to King James.	Apr.: Launched proceedings to obtain deeds to Blackfriars Gatehouse (P273). May: Prematurely mentioned as being dead in the legal case *Ostler v Heminges* (P273).	Susan de Vere Herbert's brother-in-law the Earl of Pembroke was appointed Lord Chamberlain to King James.	Ben Jonson's complete works published (S118). Susan de Vere Herbert's brother-in-law the Earl of Pembroke was appointed Lord Chamberlain to King James.
1616		Mar: signed will. Apr: died.		Jonson received a pension of 66 pounds a year and became the first Poet Laureate.
1618		Epitaph book by Richard Brathwait noted John Combe's monument at Trinity Church but not Shakespeare's.		
1619	William Jaggard published 10 Shake-speare reprints, 2 of which were falsely attributed, dedicated the book to de Vere's daughter Susan and her husband, and requested access to unprinted ShakesSpeare texts.		William Jaggard published 10 Shakespeare reprints, 2 of which were falsely attributed. He dedicated the book to Susan de Vere, and requested access to unprinted Shakespeare texts: the "fairest fruitages" and "bestow [them] how and when you list".	King James granted Southampton 1,200 pounds a year in lieu of land (S119).
1621				King James pursued a marriage alliance with Spain for his son Prince Charles. Anti-Spanish Marriage crusaders Earl of Southampton and Henry de Vere were arrested. Henry was put in jail (S119).

TABLE 1 (continued)

YEAR	SHAKESPEARE	SHAKSPERE	OXFORD	OTHER
1622	Separate publication of *Othello* (L); first new work since 1609. Peacham (Frankfurt Book Fair 1622) wrote that Oxford was at the top of the list of Elizabethan poets—Shakespeare not mentioned. The same statement was repeated in 1624 and 1634 editions.	Father's grave was dug up and moved.	His son, the 18th Earl of Oxford, went to the Tower, with threats of his execution.	18th Earl of Oxford went back into the Tower with threats of his execution.
1623	First Folio, *William Shakespeare's Comedies, Histories, and Tragedies*, published with 36 plays (18 previously unpublished) (L,P7,Pr176) and dedicated to the Earls of Montgomery and Pembroke.	Monument was erected separately from the gravestone in Trinity Church, which said "look there at the gravestone, which is all he hath writ".	Son released from the Tower.	Spanish marriage plans collapsed. 18th Earl of Oxford was released from the Tower, and a Florentine courtly correspondent noted about it: "All's well that ends well".
1624	*All's Well That Ends Well* reappeared after 20 years. *Love's Labours Won* renamed (L).		Son Henry died during a battle in the Low countries. He was 31 years old. Oxford's direct male line died out. Henry's second cousin Robert de Vere, a soldier in the Dutch army, became the new Earl. When he died in battle, his five-year-old son Aubrey became the 20th, and last, Earl of Oxford.	Death of Earl of Southampton a few days after recovering from influenza while in battle in the Low countries (L). He was 51 years old.
1630	First time Shakspere was connected to Shakespeare: In *Banquet of Jests*, Stratford-upon-Avon was said to be "a town most remarkable for the birth of Wm. Shakespeare" (Ch195).			
1632	Second Folio published (L). 21 of 30 Lyly plays also were published and contained more "excellent language" than previous versions of his plays (L).			
1634	Print debut of *Two Noble Kinsman.* There were earlier allusions to the play in literature.	Dugdale visited Stratford-upon-Avon and created a sketch of Shakespeare, from the effigy there, with a long drooping mustache and a full beard (Ch183) and holding a woolsack.		

TABLE 1 (continued)

YEAR	SHAKESPEARE	SHAKSPERE	OXFORD	OTHER
1640	John Benson's *Poems by Wil. Shakes-Speare* criticized the First Folio's preface and Droeshout's engraving of Shakespeare (Ch195). The new engraving of Shakespeare by William Marshall has an added nobleman's cape (R654) and a letter "To the Reader" with de Vere's name encrypted in it (R655). The poems and sonnets were not published again for 70 yrs.			
1649		Dr. James Cooke visited Shakspere's daughter Susanna Hall about Dr. John Hall's papers and bought 2 medical casebooks handwritten in Latin. There was no mention of any papers of her father's (Ch131). The bust at Trinity Church was re-beautified this year, according to the 1907 *Encyclopedia Britannica.*		
1664	Plays added in to the Third Folio: *Pericles, The London Prodigal, The History of Thomas, Lord Cromwell, The Tragedy of Locrine, Sir John Oldcastle, The Puritan Widow, A Yorkshire Tragedy. Pericles* not considered apochryphal (Ch68).			
1721		Effigy in Trinity Church changed to include a goatee and an upturned mustache.		
1732			Francis Peck wrote that he planned to print a 1580 comedy by de Vere and said had been an early draft of *Twelfth Night* (A154).	
1748		Repairs made to effigy.		

Journal of Scientific Exploration, Vol. 34, No. 2, pp. 351–354, 2020   0892-3310/20

New Life for Cold Fusion

HENRY H. BAUER

207 Woods Edge Court, Blacksburg, VA, USA 24060-4015
hhbauer@vt.edu

Submitted November 6, 2019; Accepted January 8, 2020; Published March 30, 2020

https://doi.org/10.31275/2020/1713

Abstract—The take-away from this discussion is that research on nuclear reactions occurring at ordinary temperatures in certain metals with electrolysis in heavy water ("cold fusion"), which has been widely denigrated for three decades as "pathological science," has now been recognized by mainstream sources as a respectable topic for further research.

Keywords: cold fusion; low-energy nuclear reactions (LENR); condensed matter

BACKGROUND

In many quarters, in most of the mass media, "cold fusion" has remained among the class of pseudo-scientific topics, analogous to perpetual-motion machines: mistakes fueled by sloppiness or wishful thinking, or perhaps deliberate hoaxes; at any rate, not to be taken seriously.

Cold fusion had made its debut in 1989 at a press conference at the University of Utah when Stanley Pons and Martin Fleischmann revealed that they had observed, in electrochemical cells with palladium (Pd) electrodes and heavy water (D_2O), the generation of heat so great that it could be attributed only to nuclear rather than chemical reactions.

A spate of hurried attempts at replication followed all over the world, often by groups with no experience in electrochemistry. They failed to confirm the claim, which was quickly labeled as "pathological science," primarily by the physics community.

Nevertheless, quite a large number of researchers, chiefly electrochemists, continued to work in the belief that Fleischmann and Pons were on to something; Fleischmann in particular was a highly respected scientist. As a result, the field came to be described not as cold fusion but as Condensed Matter Nuclear Science (CMNS) or Low-Energy Nuclear Reactions (LENR); and by 2019 the International Society of CMNS was publishing the 29th volume of its journal (iscmns.org); LENR-CANR.org boasts a library of 4,500 journal articles on the subject; and there have been more than 20 international conferences on the matter.

Nevertheless, the subject remained anathema in mainstream circles, so it was a surprise when an acknowledged mainstream source widely regarded as authoritative, *Nature* magazine, published "Revisiting the cold case of cold fusion" (Berlinguette et al., 2019) together with an Editorial preview (*Nature* Editorial, 2019) of the article.

Unsurprisingly, *Nature*'s Editorial was snarky and misleading, in asserting that "The phenomenon—even if real—seemed ephemeral and had little to no theoretical basis." In fact, Fleischmann had long been intrigued by widely acknowledged oddities in the electrolysis of aqueous solutions at Pd electrodes, and he had pointed out that sufficiently high overvoltage (away-from-equilibrium electrode-potential) would correspond to pressures of D in Pd comparable to what "hot" fusion research is aiming to achieve (Bauer, 1990). "The team found **no evidence whatsoever of cold fusion**" [emphases added]. Yet it was acknowledged that "The group was unable to attain the material conditions speculated to be most conducive to cold fusion." So the lack of evidence means nothing beyond the experimenters' failure to achieve the conditions that McKubre's group (see below) had achieved.

It is worth bearing in mind always that *Nature* (as also *Science*) suffers the self-inflicted dilemma of aiming to be both authoritative and also first with news of important advances (Bauer, 2012, pp. 67–69; Bauer, 2017, pp. 110, 162). In practice, rarely will *Nature* publish anything counter to the conventional wisdom, no matter how many well-qualified but maverick experts support the unorthodoxy (Bauer, 2017, pp. 193–194).

By contrast to *Nature*'s Editorial comment, Berlinguette et al. (2019) regarded their 4-year project as yielding useful knowledge and urged other researchers to "produce and contribute data in this intriguing parameter space. . . . the search for a reference experiment for cold fusion remains a worthy pursuit because the quest to understand and control unusual states of matter is both interesting and important." That positive conclusion may explain why it took *Nature* a year to publish the article ("Received 25 May 2018; Accepted 11 March 2019; Published online 27 May 2019"). Another inducement to publish may have been that the new research on cold fusion had been instigated by Google.

MCKUBRE'S COMMENTARY

McKubre is an electrochemist, now retired from Stanford Research Institute (SRI), who has worked on "cold fusion" almost from the beginning and has participated prominently in the associated conferences and organizations. Moreover, he had been in touch with Google and the prospective researchers when the Berlinguette project was initiated five years ago. In a Commentary in *Infinite Energy* (McKubre, 2019), McKubre points out the benefits accruing from the publication of Berlinguette et al. (2019) in *Nature*. First, that the work was stimulated by Google's recognition that the existing known sources cannot satisfy the future energy needs of Earth's growing and developing population. Second, the article confirmed one of the points McKubre's own work had established, namely that the phenomenon could be observed only when the ratio of absorbed D atoms to metal-lattice Pd atoms exceeds 0.875. Third, the very fact of publication in *Nature*, which up to now had deliberately and studiously treated the subject as beyond the pale, represents an inestimably significant breakthrough that can serve to open doors for venturesome young researchers to carry the work forward.

McKubre also makes two serious criticisms: First, the article gives a misleading view of what "cold fusion" researchers have ventured as possible mechanisms. Soon abandoned was the simplistic notion that what occurs is essentially the same in terms of fusion products as in hot fusion. Rather, its occurrence in the solid state—inside the Pd electrode—means that the palladium-metal lattice plays a crucial role.

That is why the research community adopted the name Condensed Matter Nuclear Science (see iscmns.org) to replace "cold fusion." Second, the article ignores previous work that had shown the need not only for high loading of gas into Pd but also for sufficiently high current-density applied for periods as long as several weeks, before the heat observed by Pons and Fleischmann would manifest.

The takeaway moral is that research on nuclear reactions occurring in the solid state in certain metals at ordinary temperatures, generally classed as pathological science for three decades, has been recognized as respectable for mainstream researchers, which should bring resources and general support that has been lacking up to now.

REFERENCES

Bauer, H. H. (1990). Physical interpretation of very small concentrations. *Journal of Scientific Exploration, 4*(1), 49–53.
https://www.scientificexploration.org/docs/4/jse_04_1_bauer.pdf

Bauer, H. H. (2012). *Dogmatism in science and medicine: How dominant theories monopolize research and stifle the search for truth.* McFarland.

Bauer, H. H. (2017). *Science is not what you think: How it has changed, why we can't trust it, how it can be fixed.* McFarland.

Berlinguette, C. P., Chiang, Y.-M., Munday, J. N., Schenkel, T., Fork, D. K., Koningstein, R., & Trevithick, M. D. (2019, June 6). Revisiting the cold case of cold fusion. *Nature, 570,* 45–51. https://doi.org/10.1038/s41586-019-1256-6

McKubre, M. C. H. (2019, July/August). Critique of *Nature* Perspective article on Google-sponsored Pd-D and Ni-H experiments. *Infinite Energy, 146.*

Nature Editorial. (2019, May). A Google programme failed to detect cold fusion—But is still a success. *Nature, 569,* 599–600.
https://doi.org/10.1038/d41586-019-01675-9

Journal of Scientific Exploration, Vol. 34, No. 2, pp. 355–363, 2020   0892-3310/20

ESSAY REVIEW

The Star Gate Archives, Volumes 1–4, Reports of the United States Government Sponsored Psi Program, 1972–1995 edited by Edwin C. May and Sonali Bhatt Marwaha. McFarland, 2018–2019. 2,342 pp. (all 4 volumes). $95 for each volume (paperback). ISBNs: 978-1-4766-6752-2; 978-1-4766-6753-9; 978-1-4766-6754-6; 978-1-4766-6755-3. Vol. 1: Remote Viewing, 1972–1984, 546 pp. Vol. 2: Remote Viewing, 1985–1995, 614 pp. Vol. 3: Psychokinesis, 467 pp. Vol. 4: Operational Remote Viewing: Memorandums and Reports, 715 pp.

REVIEWED BY DAMIEN BRODERICK

San Antonio, Texas
thespike@satx.rr.com

https://doi.org/10.31275/2020/1631

For your consideration, two fragments of Twilight history (as Rod Serling might have put it): a dimension as time-strung as eternity, unnerving as a grating laugh at three a.m. on a dark, chilly morning.

One: In 1946, a would-be suicide named George B. J. Stewart attracted the interest of a beefy, bearded, wingless angel named Santa Claus, and discovered how to shift into mirror universes. The post-Second World War US Congress quickly established a research center to contact other angels, especially those with working wings, and subsidized the program until 1974, when President Nixon's resignation caused funding to dry up. Despite top-secret classification masking the CLARENCE program, Stewart is rumored to be alive and still active at the North Pole at the age of 111.

Two: In 1972, three Scientologists and the brother-in law of the third best chess grandmaster in history were invited by the US military to launch what would become a $19.933 million program devoted to

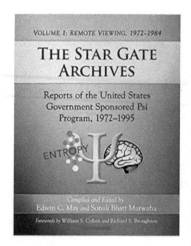

VOLUME I: REMOTE VIEWING, 1972-1984

THE STAR GATE
ARCHIVES

Reports of the United States
Government Sponsored Psi
Program, 1972-1995

ENTROPY Ψ

Compiled and Edited by
Edwin C. May and Sonali Bhatt Marwaha
Forewords by William S. Cohen and Richard S. Broughton

research into psychic powers. The initial emphasis was operational, with trained clairvoyants casting their attention into far lands and even the future. Many branches of the intelligence community sought specific double- or triple-blind tasking, alarmed by rumors that the Soviets were making advances in this domain. Despite popular rumors, the CIA was not heavily involved; the major funder was DIA (Defense Intelligence Agency). Along with NASA, DARPA, US Army Medical Research and Development Command, Foreign Technology Division, and others, DIA repeatedly contracted for this espionage methodology.

Which, if either, of these ludicrous accounts is true? Well, it turns out that CLARENCE is merely a tall story (one I just concocted). By contrast, military research programs into psychic phenomena became public after long-hidden secret documents surfaced. Most recently, four immense volumes have been published by McFarland—dubbed collectively *The Star Gate Archives*—providing an opportunity to track government-funded scientific research into psi (purported mental abilities able to reach beyond limits established by canonical sciences). Despite those limits, for two decades the science edge of the program was situated on the West Coast at Stanford Research Institute (SRI) and then Science Applications International Corporation (SAIC). A 2017 summary paper states:

> In July 1972, Russell Targ, as principal investigator, submitted a grant application on Research on Techniques to Enhance Extraordinary Human Perception to the Jet Propulsion Laboratory, NASA, with Dr. Harold Puthoff as co-investigator. This started the SRI program in psi research, which eventually closed in 1995 at SAIC. (Marwaha & May, 2017)

Its two most effective founding viewers were Ingo Swann and Pat Price, now deceased, both devotees of L. Ron Hubbard's Scientology cult.

For internal security reasons, the success or failure of individual efforts were rarely revealed. But since the psi operatives were sometimes called back for further clandestine tasking, it seems evident that the results were often sufficiently effective and accurate in support of more conventional intelligence activities. There's ample evidence for this in the various volumes. One 1984 letter of appreciation from the Deputy Director for Communications Security at the National Security Agency is displayed in Volume 4.

> To: Commander U.S. Army Intelligence and Security Command
> ATTN: Chief, Security Systems Planning Division [. . .]
>
> 1. We wish to express our thanks and appreciation for your outstanding support to [REDACTED] At our SG1A request, you were able to provide immediate, specific information, some of which was later confirmed or complemented by information from other sources. Overall, your support considerably enhanced the scope of the project and resulted in tangible success and genuine impact on U.S. national security.
> 2. [REDACTED] has received attention at the highest levels of the U.S. government. Your contribution is considered significant, and will be used for future considerations as it has been in the past. [. . .]

Despite such cagey testimonials, the program was formally closed down in 1995 when, after a rudimentary examination, the CIA deemed the results insufficiently reliable. See deft individual summaries of these four volumes by Mörck (2018, 2019, 2020a, 2020b).

Most of the operational applications (or, more candidly, psychic spying) were conducted in a rather shabby building on the grounds of Fort Meade, Maryland. These efforts were scrutinized, approved, and improved by authorized and usually disinterested specialists including a Scientific Oversight Committee (1986–1995), an Institutional Review Board, and a Department of Defense Policy Oversight Committee. A notable advocate of the project was Dr. Jack Vorona, then Deputy Director Science and Technology, Defense Intelligence Agency. Here is a startling summary from the fourth volume:

> Between the SRI and the remote viewing (RV) operations
> group at Ft. Meade, a total of 504 separate missions were
> tasked by a variety of agencies that required 2,865 individual
> remote viewings to accomplish the stated missions. Of the 19
> client agencies from 1973–1995, 17 were returning customers.

So how is this seeming craziness possible? Is it more believable than imagining a military research study of Santa Claus building toys with his elves in an icy workshop? It's a matter of credibility, but of a special kind. Many established scientists do not find psi believable because it's, well, darn it, just too gosh-heck *un*believable. No need to look at the data, at the purported empirical evidence. Do you need to *test* the claims of flat-earthers and foil-hat schizophrenics? Psi has to be just as fraudulent, critics assert, or carelessly gathered and incorrectly analyzed.

Regard the standard skeptical reasoning in action. Recently, a notable academic journal published "The Experimental Evidence for Parapsychological Phenomena: A Review" by Lund University's Etzel Cardeña, Thorsen Professor of Psychology (Cardeña, 2018). In June 2019, two US psychologists rebuked Cardeña, explaining how they just know in their rigorous bones that such psi capacities are non-existent. A. S. Reber and J. E. Alcock published "Searching for the Impossible: Parapsychology's Elusive Quest" in the same journal (Reber & Alcock, 2019a). Several months later, a slightly revised version appeared in the *Skeptical Inquirer*, where Alcock and Reber stated their approach even more firmly:

> Recently, *American Psychologist* published a review of the evi-
> dence for parapsychology that supported the general claims of
> *psi* (the umbrella term often used for anomalous or paranormal
> phenomena). We present an opposing perspective and a broad-
> based critique of the entire parapsychology enterprise. Our
> position is straightforward. Claims made by parapsychologists
> cannot be true. The effects reported can have no ontological
> status; the data have no existential value. (Reber & Alcock, 2019b)

Do they know this because of their scrupulous study of those claims and experimental data? No, it turns out. In their revision, they state baldly:

> We did not examine the data for psi, to the consternation of the parapsychologist who was one of the reviewers. Our reason was simple: the data are irrelevant. We used a classic rhetorical device . . . a form of hyperbole so extreme that it is, in effect, impossible. Ours was 'pigs cannot fly'—hence data that show they can are the result of flawed methodology, weak controls, inappropriate data analysis, or fraud. [Italics added] (Reber & Alcock, 2019b)

They were hardly the first to make this eyes-tight-closed confession. Famous science writer Isaac Asimov rejected psi, saying "If you came to me . . . and demonstrated [psychic phenomena] I would probably proceed to disbelieve my eyes. Sorry . . . " It's a common assessment, so it seems hard to believe that psi (although not Santa Claus) should be put to the test with government approval and funding. When it was shut down after 23 years, though, the justification was not "It's *impossible!*" Rather, former US Senator William S. Cohen—for ten years the ranking member of the Senate Intelligence Oversight Committee—notes that his initial "high bar of doubt began to descend as I listened to and observed the participants in the Star Gate program" (Foreword, in all 4 volumes). He concludes: "I believe it was a mistake for us to abandon the effort . . . " Insiders have told me that the closure was driven not by *failures* of the program but by its frightening degree of *success*. Certain influential military and political figures were convinced that such remote viewing successes had to be due to . . . *the influence of Satan*. But in general it was post-Cold War budget cuts and downsizing—the "peace dividend"—that really spelled its doom.

So how successful was remote viewing, done right? The Introduction to Volume 1 notes:

> On 5 October 1983, Secretary of the Army John O. Marsh, Jr., was briefed by LTC [Lieutenant colonel] Brian Buzby, project manager, INSCOM Center Lane. Buzby reported that about

350 missions out of 700 (50%) were deemed to possess intelligence value, and 85% showed positive evidence for remote
viewing . . . a CL [Center Lane] 1990 analysis of forty-one
evaluated operational remote viewings indicates that 41.5%
of the remote viewings had intelligence value. . . . Considering the nature of remote viewing these numbers are truly
remarkable.

Similarly, a 1983 Grill Flame report states:

evaluation by appropriate intelligence community specialists
indicates that a remote viewer is able by this process to generate useful data corroborated by other intelligence data. As
is generally true with other human sources, the information
is fragmentary and imperfect, and therefore should not be
relied on alone but is best utilized in conjunction with other
resources.

When the documentation of the two decades of research and
practical remote viewing was opened up by the CIA at the start of the
21st century, the declassified material was indigestible, unordered,
impossible for any but the most deeply embedded to comprehend.
Edwin May, long-term director of the program, with his associate
Sonali Bhatt Marwaha (who did most of the document sorting and
scanning scutwork over five years), organized this hoard into a genuine
archive preserving the history of this unlikely program, providing notes,
bibliographies, appendices, glossary, and indexes. Here is the bottom
line, spelled out by Dr. Richard Broughton in a second Foreword:

the most dramatic realization to emerge from Star Gate is
that psi could be useful. . . . When intelligence agencies need
information about a situation . . . they will deploy all the tools
at their disposal. . . . Psi does not enter the picture as some
sort of magic power that will give them the answer. It is just
one more of the tools that can be deployed. . . . The take-
home message is that *psi isn't magic.*

Not only is psi not magic nor diabolic intervention, as the Editors note,

> Right from its inception, the SRI–SAIC program has taken a physicalist position [that is, based on known sensory aspects of perception] in the exploration of precognition, clairvoyance, and psychokinesis—primarily a physics, engineering, and cognitive science approach. Although the SRI team explored psychological correlates such as personality (which did not lead them far), there is absolutely no mention of terms such as consciousness (except stray references to consciousness as a general term), non-local consciousness, spirituality, dualism, or religion in the SRI–SAIC reports. (Marwaha & May, 2017)

Little wonder that not only hard-shell scientists repudiate its findings (almost always without reading them); so too do many of the die-hard mystics, reincarnation mediums, prosperity gospel touters, and other devotees of superstition.

Here is a small irony of history that added to the disapproval of those who find the program's last codename cheap, derivative, and comic-bookish. In reality, the science fiction movie *Stargate* came out in 1994 and the TV series in 1997. Both had been preceded by the renaming of the US psi program to Star Gate in 1991. But luckily, these volumes are of more than antiquarian interest. After the multiply-named program was defunded and shuttered, May and some of his colleagues continued developing a theoretical attack on the puzzles of psi at the Laboratories for Fundamental Research in Palo Alto, California, summarized here as well.

Their prime model is DAT—decision augmentation theory—in which an unconscious awareness of future events can bias choices in the present. If a drunken driver is on a ragged course to smash into you from a side street, a psi warning might provide an urgent prompt to slow down or change lanes. Part of that informational schema predicts constraints mapped by entropy gradients, where a future "target" becomes, so to speak, more or less vividly detectable according to how much its elements change. It's easier to detect a nuclear weapon explosion or a deadly car crash than a cute snoozing bunny or a restful lake.

Crammed with official and long-classified reports on the program, some illustrated, some with handy charts, these four books range from 466 to 715 big pages. Two volumes focus on remote viewing, a third on causal psi, aka psychokinesis—for which they found no strong evidence—and a final, portly, behind-the-scenes collection draws upon 11,067 official reports on studies and operations. The first three are hefty, data-choked, double-columned, while the fourth fills each broad page with often name-redacted scans of memoranda, reports, and customer evaluations (yes, hundreds of tables, figures, and equations—a manager's dream.)

In short, they are not meant for light gym or beach reading. But they might change some skeptical minds, and offer hints of paths to a genuine science and technology of these apparently informational but rare abilities. However, it is all too likely that if a major theoretical breakthrough incorporates precognition, the work of long-ridiculed psi researchers, not least those from the Star Gate program, will be entirely ignored by the new Nobel Prize candidates.

If there is one drawback to these useful compendiums, it is the tightly crammed spines of their large, heavy paperbacks. Without powerful psychokinetic assistance (which, remember, the Star Gate scientists say does not actually exist), you can't leave the book open on the page without it springing shut. Even holding it down in a muscular wrestling deathlock does not bring it into submission, because on many pages the text at the right or left inner margin vanishes into the spine. You can break the back of the books in numerous places, but that is not recommended. Luckily, McFarland also offers an e-book option, where the small print can be expanded to improve readability, and the spinal crushing is no more. I recommend the e-book editions (which are also about half the price) for home or office reading, and leave the heavy-duty paper volumes for libraries—which should certainly accession this remarkable quartet.

REFERENCES

Cardeña, E. (2018). The experimental evidence for parapsychological phenomena: A review. *American Psychologist, 7395*(5), 663–677. https://doi.org/10.1037/amp0000236

Marwaha, S. B., & May, E. C. (2017). *The Star Gate archives: Reports of the U.S. Government sponsored psi program—1972–1995, an overview.* The Laboratories for Fundamental Research. [Updated edition of paper presented at the 60th Annual Convention of the Parapsychological Association, Athens, Greece, July 20–23, 2017.] https://www.academia.edu/38006378/the_star_gate_archives_reports_of_the_US_government_sponsored_psi_program_1972-1995._an_overview

Mörck, N. (2018). *The Star Gate archives* (volume 1) [book review]. *Journal of Scientific Exploration, 32*(4), 773–780. https://www.scientificexploration.org/docs/32/jse_32_4_Morck.pdf

Mörck, N. (2019). *The Star Gate archives* (volume 2) [book review]. *Journal of Scientific Exploration, 33*(2), 299–308. https://www.scientificexploration.org/docs/33/jse_33_2_Morck_on_Stargate_2.pdf

Mörck, N. (2020a). *The Star Gate archives* (volume 3) [book review]. *Journal of Scientific Exploration, 34*(1), 124–136. https://www.scientificexploration.org/docs/34/JSE_34_1_Morck_Review.pdf

Mörck, N. (2020b). *The Star Gate archives* (volume 4) [book review]. *Journal of Scientific Exploration, 34,* forthcoming.

Reber, A. S., & Alcock, J. E. (2019a). Searching for the impossible: Parapsychology's elusive quest. *American Psychologist, 75*(3), 391–399. https://doi.org/10.1037/amp0000486

Reber, A. S., & Alcock, J. E. (2019b). Why parapsychological claims cannot be true. *Skeptical Inquirer, 43*(4), 8–10. https://skepticalinquirer.org/2019/07/why-parapsychological-claims-cannot-be-true/

Journal of Scientific Exploration, Vol. 34, No. 2, pp. 364–372, 2020   0892-3310/20

ESSAY REVIEW

Scientocracy: Why a Science Court Is Needed

Scientocracy: The Tangled Web of Public Science and Public Policy edited by Patrick J. Michaels and Terence Kealey. Cato Institute, 2019. 368 pp. $19.95 (paperback). ISBN 978-1948647496.

Reviewed by Henry H. Bauer

Virginia Polytechnic Institute & State University
hhbauer@vt.edu www.henryhbauer.homestead.com

https://doi.org/10.31275/2020/1745

The Introduction is spot on: "Science *can* be a force for good, and it has enhanced our lives in countless ways, but even a cursory look at the 20th century shows that what passes for science can be detrimental" (p. 1).

Ten of the eleven chapters of this book are comprehensively documented case studies demonstrating in each instance that the pertinent public policies were based on or justified by supposedly scientific understanding when in reality the so-called science was quite inadequate to support those policies; and moreover was severely biased by conflicts of interest and vested interests of the panels and advisory committees given the responsibility for assessing the actual state of the relevant scientific knowledge. These case studies are valuable, including as resources for other scholars, and it is regrettable that much of the book would have benefited from better copyediting; for instance, on page 55, "life expectancies . . . have continued to rise . . . by three months for every year lived"—what does that mean? The reference (#48) given as source doesn't help because its URL link doesn't work. And it is also not helpful to learn that "Western diet had the dual effect of both stimulating and damaging our health" (p. 55), particularly since the cited reference (#49) says nothing about "Western diet". It is also annoying that the figures are too small, and that the colors relied on for making distinctions are too faint and indistinct in several figures.

The book does illustrate quite convincingly that the nature of contemporary scientific activity is nothing like the traditional, conventional view of science as a disinterested activity delivering public goods. As discussed in comprehensive detail in *Science Is Not What You Think* (Bauer, 2017a), among the salient factors for this difference are the rat race for obtaining funds for research and the fact that peer review serves only to entrench whatever the consensus is among the dominant cliques in each scientific specialty.

Chapter 1 of the book is colored by the Cato Institute's libertarian ideology, seeking to make government funding of research the culprit for the dysfunctional state of affairs; and that view is parroted to some extent in various ways in later chapters. The argument is made by data indicating that governmental funding of research did not increase national GNP per capita (p. 26). But that is not an appropriate measure of what is good for the population as a whole. GNP in the US, for example, would be lower if the healthcare system cost less while delivering better health outcomes, as is the case in Canada, Australia, and many European countries where government manages healthcare more directly and firmly.

Although it is certainly true that the corruption of scientific activity began to increase when government funding for research increased enormously, namely after the end of World War II, government funding is not uniquely or primarily responsible for what is wrong with science nowadays. The fundamental problem is that so many areas of research now require more funding than individual universities are able or willing to supply. Researchers must therefore obtain resources through their own efforts, and the old saying is perfectly applicable: "Those who pay the piper, call the tunes." That is certainly true for industrial funding of academic research as much as it is for industrial funding of in-house research and as it is for funding by government agencies; indeed, it is true for much of the funding from private charitable foundations, which naturally and quite properly support research that is likely to promote the causes the foundations are set up to advance.

Chapter 1 is unusually good, however, in describing how science and technology are related (or not; see also pp. 161–162); and it is also unusually good in describing the many mis- and ab-uses of statistical analysis, one consequence being that fields relying on statistical analysis suffer pervasively from the currently deplored "crisis of (ir) reproducibility" (pp. 28–29).

Chapter 2 documents the sad story of misleading advice about risks allegedly associated with various forms of dietary fat. There is actually no convincing evidence that saturated fat in the diet is harmful, and it is simply wrong to claim that high blood levels of cholesterol (or one of its forms) cause cardiovascular problems (e.g., Ravnskov, 2000; Kendrick, 2008, 2014). It took 60 years for misguided official warnings against cholesterol-rich foods to cease, illustrating a common dysfunction (p. 45): Once official advice has been issued, even when based on inadequate evidence, it is a Herculean task to have it modified even as convincing evidence mounts; many practicing doctors continue to believe these falsehoods (p. 46). Figure 2.1 (p. 40) indicates that mortality from strokes is not caused by atherosclerotic heart disease since the incidences changed differently over the years (pp. 52, 53).

It irritates me greatly when a book gets simple arithmetic wrong. Chapter 2 states (p. 54) that the change from 2.6% to 6% is "more marked" than that from 13.4% to 30.9%, yet in both cases the ratio is the same, 2.3076 and 2.3060, to be pedantically accurate. Such careless innumeracy makes questionable everything in this chapter wherein data matter.

Chapter 3 demolishes the still-prevalent myth that public health would be served by restricting the intake of salt; and it describes how the system of committees and official agencies kept the myth hegemonic, with facts ignored or distorted; it was necessary only "to convince key government officials and the public" (p. 97). Blood pressure is a biomarker for salt intake, but that does not make it a valid measure of overall health outcomes (p. 107). Harmful misuse of biomarkers is widespread (Institute of Medicine, 2010, 2011).

Chapter 4 points out that hysteria over drug abuse has had the very harmful consequence of denying pain-alleviating medication to some who genuinely need such relief: There was no correlation between number of prescriptions written in a given region and the number of addicts there (p. 132 f.); indeed, all the popular beliefs about the opioid epidemic are wrong (p. 138). Data on potency (p. 117) may be of general interest: heroin and methadone, 2.5 times as potent as morphine, hydromorphone about 6 times as potent as morphine, fentanyl 50 times as potent as morphine; while oral codeine is only 1/6 as potent as morphine.

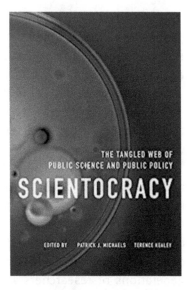

THE TANGLED WEB OF
PUBLIC SCIENCE AND PUBLIC POLICY

SCIENTOCRACY

EDITED BY   PATRICK J. MICHAELS   TERENCE KEALEY

Chapter 5 describes how bureaucratic entrenchment of beliefs about the great dangers of marijuana, psilocybin, LSD, and similar drugs has dysfunctionally prevented research that could well establish useful medical applications for these and related substances. "Ecstasy" (MDMA) had been patented in 1914 and used for a long time in psychotherapy (p. 153 ff.), with particular success in cases of PTSD (p. 159). Prohibition is no substitute for sensible regulation (p. 160).

Chapter 6 is a largely well-founded tirade deploring how bureaucracy, owing chiefly to government actions, works against useful medical innovation. But the libertarian bias for private as opposed to government funding is pervasively overt in this chapter, and contrary to actual experience with respect to academic research (p. 165); the bias goes so far as to describe as "reliable medical information" what drug representatives convey to doctors (pp. 179–180). The chapter is also seriously wrong on one very important point: in welcoming, as an example of desirable non-government–influenced medical innovation, the introduction of statin drugs (p. 168), which in reality cause demonstrable harm by weakening the body's energy-producing mechanisms (e.g., Langsjoen & Langsjoen, 2008; Langsjoen et al., 2008; Hansen et al., 2005; Anonymous, 2010; de Lorgeril & Rabaeus, 2015; Rabaeus et al., 2017) and whose supposed benefit is based on the mistaken view (Ravnskov, 2000; Kendrick, 2008, 2014) that high cholesterol levels in the blood constitute cardiovascular disease. The building up of plaque in the arteries is initiated by inflammation or physical damage, and some of the occasional benefit attributed to statins can be explained by their somewhat anti-inflammatory properties. That "80 percent of . . . drug approvals arose solely from . . . private industry" (p. 167) reflects not service to public health but rather the marketing of me-too modifications and substances, like statins, of at-best–doubtful public benefit (Moore, 1995, 1998; Angell, 2004; Goozner, 2004; Moynihan & Cassels, 2005; Brownlee, 2007; Greene, 2007; Petersen, 2009; Healy, 2012; Goldacre, 2013; Gøtzsche, 2013).

Chapter 7 demonstrates that official regulation of carcinogens and other chemicals is based on a fallacy, namely that the risk of harm is linearly proportional to the exposure dose. Amply documented is how this came about, in part through deliberate distortion of evidence by self-interested people and groups; and it illustrates once more how such a fallacy can persist for a long time—several generations of researchers and practitioners (p. 189). In reality, the pertinent data and evidence show beyond doubt that there is no harm below a certain threshold, or in many cases and even more strikingly that substances harmful at high doses may actually be beneficial at very low exposures. That phenomenon, hormesis, though perhaps surprising at first mention, has an entirely conventional and logical basis: The immune system detects potential harm and is activated beyond its normal resting state, with beneficial side effects. The author of this chapter, Edward Calabrese, is also the scientist whose work established hormesis as a general phenomenon, and his work has been supported by government funding (p. 207). This chapter alone is worth the price of the whole book.

Chapter 8 builds on the knowledge conveyed by chapter 7 to show how ignorance of hormesis with respect to exposure to radiation has been harmful in the long-drawn-out battle over whether to permit the mining of uranium in Virginia. Illustrated is that the National Research Council cannot be relied on for an accurate, impartial assessment.

Chapter 9 relates the continuing battle over potential mining of mineral deposits in Alaska. Here the salient factor is not so much bureaucratic reliance on faulty science as bureaucratic arrogance in abusing current legislation by even preempting scientific assessments.

Chapter 10 focuses on some aspects of the mistaken view that carbon dioxide is the primary and harmful cause of global warming;

and it shows that the computer models on which all climate-change hysteria is based are simply wrong; they are "tuned" subjectively (p. 250), in other words rely on fudge factors. Focusing on highly technical details, this chapter could have been more readily accessible to most readers if it had also emphasized the actual historical data of periodic ice ages separated by much warmer periods, as well as the data comparing historical temperatures with contemporaneous levels of atmospheric carbon dioxide (Bauer, 2017b). Nevertheless, this is a clear demonstration that existing courts cannot deliver properly informed judgments when questions of disputed scientific understanding are at issue (p. 240).

Finally, chapter 11 details the excruciatingly bad statistics applied in the regulation of fine particulate matter in the atmosphere. This willfully ignorant and shockingly incompetent resort supposedly to statistical science applies also to earlier chapters. Almost all or perhaps even all of the dysfunctional official advice about nutrition in particular and health in general transgresses perhaps the most important single thing about statistical analysis; namely, that correlation or association is no proof of causation. Beyond that, the attempt to detect single causes for the effects of diet or of the environment seems wrong-headed a priori: Essentially innumerable possible influences exist and there is no satisfactory way to control for the possible influences of factors other than the one of specific interest in any given research. Given that the statistical analyses do not hold water, I wondered about animal studies of harm from inhalation of fine particulate matter and was disappointed to find no discussion of the host of existing studies.

This book demonstrates beyond any reasonable doubt that policymakers and regulating agencies cannot obtain impartial, disinterested, objective assessments of the state of scientific knowledge from existing sources. The National Academy of Sciences, the National Research Council, and all other scientific and academic organizations, be they national or international, private or public or governmental, inevitably reflect the prevailing scientific consensus, the conventional wisdom within the scientific community; and that is simply the opinions within the currently dominant clique. It is quite obviously impossible to obtain impartial, disinterested assessments from the putative experts in any given field since the conventional wisdom in that field is synonymous with the views of those experts.

Law-makers and policy-makers and regulators do, however, sorely need impartial, disinterested, objective assessments of the state of scientific knowledge on any matter pertinent to public policies. The suggestion (p. 59) that investigative journalists can deliver these goods is whistling in the wind, since their published findings have no power to force compliance with impartial fact. To deliver impartial, disinterested judgments *and to enforce appropriate compliance* on controversial matters, society developed the system in which courts supervised by disinterested judges allow opposing points of view to be presented and argued, under cross-examination and with the assistance of pertinent witnesses. The necessary decision is arrived at either by a single judge, or by a panel of judges, or by a jury of people selected without vested interest in the result. The same sort of arrangement, in the form of a specifically Science Court, seems to be the only conceivable way in which society could have the benefit of truly impartial assessments of contemporary scientific understanding.

The concept of a specifically Science Court dates back at least half a century. Kantrowitz (1967) suggested that an "Institute for Scientific Judgment" was needed as policymakers were being exposed to sharply differing scientific opinions about the potential safety of atomic reactors for generating power for general civilian use. Over the years, a number of discussions ensued about the concept, soon described as a Science Court. A further justification for such an Institution lies in the difficulties that the civil court system faces when matters of scientific knowledge and understanding are at issue, which the courts are simply not equipped to handle (Jurs, 2010); for instance, courts need to determine whether witnesses called as experts by opposing parties genuinely deserve to be regarded as expert, which implies impartial. Moreover, federal regulators and officials can ignore findings by a civil court on matters of scientific understanding, as happened with keeping marijuana as a Schedule 1 drug (pp. 144–145). The many points to be considered in the possible establishment of a Science Court have been discussed in the previously cited book by Bauer (2017a, Chapter 12).

References

Angell, M. (2004). The *truth about the drug companies: How they deceive us and what to do about it.* Random House.

Anonymous. (2010, October 18). Do statins have a role in primary prevention? An update. *Therapeutics Letter, 77.* https://www.ti.ubc.ca/2010/10/18/do-statins-have-a-role-in-primary-prevention-an-update/

Bauer, H. H. (2017a). *Science is not what you think: How it has changed, why we can't trust it, how it can be fixed.* McFarland.

Bauer, H. H. (2017b). Climate-change facts: Temperature is not determined by carbon dioxide. https://scimedskeptic.wordpress.com/2017/05/02/climate-change-facts-temperature-is-not-determined-by-carbon-dioxide

Brownlee, S. (2007). *Overtreated: Why too much medicine is making us sicker and poorer.* Bloomsbury.

de Lorgeril, M., & Rabaeus, M. (2015). Beyond confusion and controversy, can we evaluate the real efficacy and safety of cholesterol-lowering with statins? *Journal of Controversies in Biomedical Research, 1*(1), 67–92. https://doi.org/10.15586/jcbmr.2015.11

Goldacre, B. (2013). *Bad pharma: How drug companies mislead doctors and harm patients.* Faber & Faber.

Goozner, M. (2004). *The $800 million pill: The truth behind the cost of new drugs.* University of California Press.

Gøtzsche, P. C. (2013). *Deadly medicines and organised crime: How big pharma has corrupted healthcare.* Radcliffe.

Greene, J. A. (2007). *Prescribing by numbers: Drugs and the definition of disease.* Johns Hopkins University Press.

Hansen K. E., Hildebrand, J. P., Ferguson, E. E., & Stein, J. H. (2005). Outcomes in 45 patients with statin-associated myopathy. *Archives of Internal Medicine, 165*(22), 2671–2676. https://doi.org/10.1001/archinte.165.22.2671

Healy, D. (2012). *Pharmageddon.* University of California Press.

Institute of Medicine. (2010). *Evaluation of biomarkers and surrogate endpoints in chronic disease* edited by C. M. Micheel & J. R. Ball. National Academies Press. https://doi.org/10.17226/12869

Institute of Medicine. (2011). *Perspectives on biomarker and surrogate endpoint evaluation: Discussion forum summary.* A. Mack, E. Balogh, & C. Micheel, rapporteurs. National Academies Press. https://doi.org/10.17226/13038

Jurs, A. W. (2010). Science court: Past proposals, current considerations, and a suggested structure. Drake University Legal Studies Research Paper Series, Research Paper No. 11-06; *Virginia Journal of Law and Technology 15*(1). https://papers.ssrn.com/sol3/papers.cfm?abstract_id=1561724

Kantrowitz, A. (1967). Proposal for an institution for scientific judgment. *Science, 156*(3776), 763–764. https://doi.org/10.1126/science.156.3776.763

Kendrick, M. (2008). *The great cholesterol con: The truth about what really causes heart disease and how to avoid it.*. John Blake. [Updated from 2007 1st ed.]

Kendrick, M. (2014). *Doctoring data: How to sort out medical advice from medical nonsense.* Columbus Publishing.

Langsjoen, P. H., & Langsjoen, A. M. (2008, December 16). The clinical use of HMG CoA-reductase inhibitors and the associated depletion of coenzyme Q10. A review of animal and human publications. *BioFactors, 18*, 101–111. https://doi.org/10.1002/biof.5520180212

Langsjoen, P. H., Langsjoen, J. O., Langsjoen, A. M., & Lucas, L. A. (2008, December 19). Treatment of statin adverse effects with supplemental Coenzyme Q10 and statin drug discontinuation. *BioFactors, 25*, 147–152. https://doi.org/10.1002/biof.5520250116

Moore, T. J. (1995). *Deadly medicine: Why tens of thousands of heart patients died in America's worst drug disaster.* Simon & Schuster.

Moore, T. J. (1998). *Prescription for disaster: The hidden dangers in your medicine cabinet.* Simon & Schuster.

Moynihan, R., & Cassels, A. (2005). *Selling sickness: How the world's biggest pharmaceutical companies are turning us all into patients.* Nation Books.

Petersen, M. (2009). *Our daily meds: How the pharmaceutical companies transformed themselves into slick marketing machines and hooked the nation on prescription drugs.* Picador (Pan Macmillan).

Rabaeus, M., Nguyen, P. V., & de Lorgeril, M. (2017). Recent flaws in evidence-based medicine: Statin effects in primary prevention and consequences of suspending the treatment. *Journal of Controversies in Biomedical Research, 3*(1), 1–10. https://doi.org/10.15586/jcbmr.2017.18

Ravnskov, U. (2000). *The cholesterol myths.* New Trends Publishing.

Journal of Scientific Exploration, Vol. 34, No. 2, pp. 373–381, 2020   0892-3310/20

BOOK REVIEW

Greening the Paranormal: Exploring the Ecology of Extraordinary Experience edited by Jack Hunter, Foreword by Paul Devereux. August Night Press, 2019. 332 pp. $19.99 (paperback). ISBN 978-1-786771094.

REVIEWED BY SHARON HEWITT RAWLETTE

sharon.rawlette@gmail.com

https://doi.org/10.31275/2020/1703
Creative Commons License CC-BY-NC

Far from a dispassionate survey of the intersection between ecology and parapsychology, Jack Hunter's recent anthology *Greening the Paranormal* is a collection of the deeply personal insights and discipline-defying questions that have arisen from the contributors' lived contact with some of the strangest aspects of the natural world.

From the very first page of Paul Devereux's Foreword, we are confronted with the inexplicably extraordinary: Devereux's sighting of a "green man" at the fork of a road in the Irish countryside. "Suddenly, standing on the grass, there was a figure, between two and three feet tall," writes Devereux.

> It was anthropomorphic and fully three-dimensional. . . . It had sprung into appearance out of nowhere, and it caught my wife's and my own transfixed attentions simultaneously. The figure was comprised of a jumble of very dark green tones, as if composed of a tight, dense tangle of foliage. . . . It presented a distinctly forbidding appearance. As we crawled past in our car, the figure started to turn its head in our direction, but then vanished. (pp. xi–xii)

Devereux, it happens, was well acquainted with ancient folklore's references to the "green man," but he admits that, until that moment, he had always considered those stories to be the product of gullibility and superstition. Not so afterward! "[G]rasping the ecological dimensions of what our culture calls the 'paranormal,'" he says, ". . . will require us to re-acquaint ourselves with some aspects of the worldviews of earlier and indigenous peoples" (p. xii).

This sentence serves as a fairly accurate summary of the thrust of the entire anthology, if re-acquaintance is understood in the sense of firsthand experience. As Hunter points out in his introductory chapter, the paranormal is actually *normal* within the context of the natural world, and if modern, industrialized society so rarely experiences the paranormal, it is because we also so rarely experience nature—that is, a world unmanipulated by human design. In fact, Hunter suggests that our society's rejection of the "paranormal" stems from the same faulty ontological assumptions that have caused the ecological crisis in which we find ourselves: the assumptions that only what is material is real, and that only what is human is valuable. Rejection of nature and rejection of the paranormal apparently go hand in hand. And so, it comes to seem in this anthology, may their recovery.

Winding through several of the essays, we find the theme that humanity has experienced its separation from the natural world as a "primordial spiritual trauma" (p. 50). We carry within us a deep wound that comes from having been torn away from our true home and our true selves. Contributor Maya Ward suggests that the depth of this primordial trauma has caused us to push away our ability to feel, leaving us with a huge backlog of unprocessed grief.

> [A]nthropocentrism could be seen as a trauma response developed over millennia but originating during the pro-found rupture of people from place that happened . . . when we moved from hunter–gatherer communities, where humans were just one creature among a society of equally sentient creatures, to farmers, where plants and animals were 'cultivated'. (p. 156)

Contributor Nancy Wissers points out that what remaining con-tacts we do feel coming from the wider earth community are now difficult for us to recognize. Wissers writes that many people, including herself, have experienced communication with the earth, but they have rarely understood it in these terms. She writes,

> Since we have little context for something from the outside touching the self without words, other than those provided by religion and culture, someone knowing only the Western worldview is likely to conceive of these golden moments as internal events of the self, or perhaps even as the presence of God. (p. 75)

Nevertheless, she argues, these are the kinds of experiences that indigenous people are talking about when they refer to being contacted by spirits (p. 74), and we still have access to these experiences, if we can recognize them for what they are.

At the same time, contributor Lance M. Foster suggests that, if our psychic connection to the natural world has been largely severed, this may have been a protective move on the part of that world, desiring to shield itself from manipulations by humankind—especially in its modern, scientific incarnation. Foster points out that, while science demands proof of unusual phenomena, it never actually stops with proof, instead always barreling ahead to develop technical (i.e. manipulative) applications of its knowledge. In fact, I would add that the quintessential way to "prove" things scientifically is to have already established some measure of control over them, which allows one to make them appear at will in the environment of the laboratory.

Foster notes that the indigenous have a very different way of responding to strange encounters, such as those with "[g]iants, little people, animals from ancient times, underwater beings, ghosts, bigfoot, sentient plants and places, things without names" (p. 91). Their response is to acknowledge these beings and respectfully leave them alone, something it seems "near impossible" for the nonindigenous to do (p. 96). In fact, contributor Cody Meyocks argues that science—a discipline based on abstracting away from anything individual or particular—has become the ultimate tool for justifying structures of political and economic domination. The deepest elements of nature can hardly be faulted if they shield themselves against its depersonalizing grasp.

Contributor Jacob W. Glazier expands on Foster's theme by using the archetype of the trickster to personify the hiding and "tricky" quality of psi. As parapsychologists well know, psi seems to flee the laboratory and to routinely frustrate, even in the field, any systematic investigation of its properties. This is one reason that contributors Elorah Fangrad, Rick Fehr, and Christopher Laursen argue for becoming psychic naturalists—studying psychic phenomena in the wild, in a participatory rather than a manipulative fashion. (They provide a detailed proposal for how they plan to pursue their study of the strange phenomena that frequently occur in a certain Ontario fishing lodge, phenomena that include anomalous figures, sounds, icy air, and poltergeist-like movements of objects.) "[M]anipulation causes the trickster to rebel," writes Glazier (p. 107), and this rebellion may be an attempt to pass on an essential message. Glazier quotes parapsychologist James E. Kennedy, who said, "The message from the trickster is that converting psi to technology is not going to happen" (p. 103).

So, given its apparent concern not to be manipulated, where does psi manifest itself most unguardedly? In "[c]ultures that are less hierarchical and more egalitarian," says Glazier, "less technological and more animistic, less ordered and more chaotic." That is, psi manifests most plainly in cultures that mirror the qualities of the trickster by "playing" with psi rather than attempting to enslave it to their purposes (p. 107).

Contributor Amba J. Sepie seconds this observation. She writes that the explanations for how indigenous people come to know the things they do about the natural world

> rely on an understanding of indigenous and traditional metaphysics and the practices which extend from these, as concretely linked to the realization that Earth and other beings are kin [meant literally], that relationships are not abstract but personal, and that consciousness is not limited to a single human mind, but rather is something all life is *internal* to. (p. 64)

In other words, "civilized" humans have been cut off from knowledge of much of nature by their unwillingness to view other beings as mental and moral equals, deserving—and rewarding—our respect and consideration.

Much of *Greening the Para-normal* is concerned with the question of how this estrangement can be repaired. Contributor Viktória Duda optimistically pro-poses that our technological out-sourcing of the processes of our own bodies and minds has built into it its own remedy. She suggests that, when virtual reality comes to be indistinguishable from "real" reality, we may realize that everything was ultimately consciousness all along. I confess, however, that I don't see how viewing the natural world as just another simulation will increase our desire to interact with it, or to acknowledge that it has its own objectively valid concerns.

A more likely remedy seems to me to lie in the fact that, as evidenced by the experiences of several contributors to this volume, our efforts to heal the rift that exists between us and the natural world are often aided by non-human minds. For instance, contributor David Luke describes his experience on a plant psychedelic, where he felt himself transformed into a thorn bush and heard all the plants around him begin laughing riotously. "[S]ince then," he says, "I have never considered ecology in quite the same way as before" (p. 182). Luke also cites two mycologists who take seriously the idea that mushrooms actually have the intention of helping us communicate with other species. And he himself has published surveys showing that encounters with plant consciousness are the most widely reported transpersonal aspect of experiences provoked by psilocybin mushrooms, ayahuasca, and the *Amanita muscaria* mushroom. In fact, 80% of psychedelic users report an increase in their subsequent interaction with nature, and more than 60% report increased concern with regard to nature. These results, Luke notes, mirror the increased ecological concern reported by near-death experiencers and those who've had UFO or alien encounters.

Moving from plants into the realm of animals, contributor Brian Taylor discusses "soul birds": the "widespread and ancient body of lore associating birds with survival beyond death and the flight of human souls" (p. 191). Taylor cites the prevalence of bird sightings around dying humans and describes his own powerful dreams and waking encounters with kingfishers, which he has come to realize have been strongly correlated with his grief over a lost loved one.

Contributor Silvia Mutterle discusses another avenue by which animal consciousness may interact with ours, describing the Wild Earth Animal Essences created by Daniel Mapel 20 years ago. In this process, the energetic vibration of the animal is apparently ceremonially "imbued" into water, carrying the species' "most outstanding characteristics and gifts" (p. 204) and allowing them to be conveyed to those exposed to this water.

I admit it was a bit difficult for me to keep an open mind about the possible efficacy of the animal essences Mutterle describes, as they are completely foreign to anything in my own experience. But I tried to keep in mind what Hunter emphasizes in his opening chapter: that we must be willing to contemplate the "weirder" kinds of paranormal phenomena, not just the "(relatively) scientifically acceptable phenomena associated with *psi*" (p. 13). Hunter mentions, among other things, UFOs, alien encounters, cryptids, and fairies. He notes that ecological themes crop up over and over again in relationship to these weirder phenomena, and it seems that we cannot study one without the other. He writes,

> I would argue that it is precisely the *most unusual* (and so least respectable)—the *Highly Strange*—accounts of para-normal experiences that we should be investigating, because they raise the most questions and challenge our established worldviews most strongly. (p. 15)

I agree with this, though I still would have liked to see a little more evidence that the Wild Earth Animal Essences are truly effective in doing what they purport to.

Speaking of weird phenomena, though, we haven't come to the end of what this book has to offer. In addition to benefiting from the contacts of animal and plant consciousnesses, it appears that our rapprochement with the natural psychic world may be aided by certain physical locations, which Hunter calls "window areas" (p. 29) and Devereux refers to as "thin" and "liminal" places. Devereux says these are "sites where 'breaking through' to otherworld realms or altered mind states . . . were felt to be more easily accomplished than other places" (p. xvii).

Contributor Mark Schroll builds on this thought by citing evidence that sacred places affect our dreams, and he proposes that rituals conducted in such places may help to heal our consciousness from its divisiveness and manipulativeness. Contributor Christine Simmonds-Moore also focuses on aspects of physical spaces and places that draw out exceptional experiences. She argues for an expanded definition of exceptional experiences that allows for experiences that "emerge as conversations between liminal people and locations that have liminal properties" (p. 126).

But no excursion into ecology would be complete without some acknowledgment that the natural world is not all fellow feeling and *kumbaya*. Animals and plants feed on one another, after all. How do we understand the predatory relationships that feature so prominently in nature, and is there any connection to parapsychology *there*?

Contributor Timothy Grieve-Carlson tackles this question in his essay on Whitley Strieber, an author who has explored the predator–prey relationship both in his fiction and his nonfiction, the latter focusing on his experiences of alien abduction. Grieve-Carlson says,

> Strieber himself has suggested that his earlier written corpus is actually the result of a lifetime of visitor experiences, sublimated and expressed through the form of the horror novel, with the visitors themselves appearing in the culturally mediated halloween fashions of the werewolf and the vampire. (p. 227)

As Grieve-Carlson brings out, Strieber's work investigates both the terrifying and mesmerizing qualities of the predator–prey relationship, as well as its potential sacredness. "This compassionate love between predator and prey in Strieber's writing is always somehow reciprocal," writes Grieve-Carlson (pp. 229–230). Despite Strieber's utter terror, he finds himself drawn to the visitors, wanting further contact, wanting "communion."

In a related vein, contributor Simon Wilson examines Paul Devereux's 1982 book *Earth Lights: Toward an Explanation of the UFO Enigma*. Wilson takes Devereux to be showing how Jung's hypothesis that flying saucers are physical manifestations of mental forces might work, and might be related to shifts in the earth's crust. But shifts in the earth's crust are not just mechanical geological happenings in Devereux's view, since the earth is a living system and these movements are actually profound changes in the earth's *body*, linked to many other changes and inextricably bound up with what's going on in the rest of the cosmos. In a UFO or Earth Lights experience, "consciousness resonates with the whole cosmos," he says (p. 173).

Might UFOs, then, be dreams of the earth—of which we are part? Our relationship to these experiences and the entities that appear to reveal themselves within these experiences is an ambiguous, almost paradoxical, one. They seem to be *other* than us and yet at the same time *inseparable* from us. Are they trying to control us? Is it our destiny to become one with them? Wilson concludes that the most inspiring picture is one in which we are meant to continue in an interactive *relationship*.

The relationship between the physical and mental aspects of strange phenomena also is explored in this anthology as it relates to cryptids. Hunter mentions in his opening chapter a suggestion made elsewhere by Joshua Cutchin: that cryptids are "wilderness poltergeists" (*wildnisgeistin*). Hunter also notes that the Loch Ness monster seems to fit aspects of a particular Hindu/Buddhist serpent deity or spirit. This raises the question of whether such "magical animals" might embody the consciousness of the places where they are found, or perhaps humans' repressed desire for interaction with them. At the same time, contributor Susan Marsh presents evidence that some cryptids appear to react to human settlement patterns in the way that flesh-and-blood animals would. I would ask: and what if they *are* flesh and blood? Does that make it impossible for these animals to *also* be the embodiment of place consciousness? Mightn't the consciousness of a place react to human settlement patterns as well?

So many invigorating questions arise from the observations contained in this volume. Ultimately, however, all of the experiences and perspectives explored in this anthology agree on at least one unequivocal point: the necessity of a model of the universe "as a living system imbued with intelligence and agency" (p. 7). Hunter reminds us in his opening chapter that "by enhancing biodiversity we are also enhancing *psychodiversity*" (p. 39). That is, we are enhancing the variety of minds working and playing together, and thus promoting the likelihood of intelligent and creative solutions to the myriad spiritual and material dilemmas that come our way.

One day, perhaps, extraordinary experiences will no longer be so extraordinary. But only, it seems, if we are willing to remember a different way of being in the world, if we are willing to open our psychic eyes and ears and be welcomed home by the many other minds who share our planet.

Journal of Scientific Exploration, Vol. 34, No. 2, pp. 382–391, 2020   0892-3310/20

BOOK REVIEW

Demons on the Couch: Spirit Possession, Exorcisms, and the DSM-5 by Michael J. Sersch. Cambridge Scholars Publishing, 2019. 95 pp. $119.95 (hardcover). ISBN 978-1527521940.

REVIEWED BY TODD HAYEN

https://doi.org/10.31275/2020/1751
Creative Commons License CC-BY-NC

Michael J. Sersch's (2019) *Demons on the Couch: Spirit Possession, Exorcisms, and the DSM-5* is an immaculately researched and referenced treatise on possession and exorcism presented through the lens of modern psychotherapy and the *DSM-5* (the diagnostic bible of the mental health field). Sersch states in his Introduction:

> In writing this book, I hope to answer why demonic possession has held a cultural fascination for over two millennia as well as how clinicians can successfully and ethically deal with patients who legitimately believe they are possessed by a spiritual force. There is also mounting evidence that integrating a patient/client's worldview into clinical practice, including their spirituality and faith practices, increases their likelihood of getting better (Lund, 2014) which is a position I am overtly advocating. (p. 5)

He also claims that he has no desire to attempt to prove or disprove spirit or demonic possession (p. 5). His approach is largely clinical and pedagogical: What does a clinician do with a patient who claims they are possessed?

Sersch divides his thesis into three sections, each dealing with a different aspect of possession and exorcism. Section 1, appropriately enough, deals with the history of spirit possession, demon possession, and different forms of exorcism. Section 2 is more clinical in its approach, going into detail on such topics as the different designations for diagnoses found in the various editions of the *Diagnostic Statistical Manual* (*DSM*) such as Multiple Personality Disorder (an older label having been replaced with Dissociate Identity Disorder in the fourth edition of the *DSM* [American Psychiatric Association, 1994]). Section 3 focuses on suggestions for the clinician: Again—How does the clinician handle patients claiming to be possessed?

Sersch's book is an excellent resource. It is quite academically robust and includes a solid reference for nearly every sentence. It has close to 20 pages of references. For a book of only 141 pages of text, that is quite an accomplishment. Sersch claims this book was originally a thesis for his Master's degree and was subsequently expanded into a more accessible read—although at times this reader still got the impression he was reading a doctoral dissertation literature review. That being said, however, the author has made it a point to interject his own personal insights from time to time in the first person, and, although a bit jarring after reading pages of academic text, it allows for a more intimate approach. I found myself wanting to hear more of Sersch's personal insights.

HISTORY

Sersch begins Section 1 on the history of possession and exorcism with a basic definition: "Possession is defined as a trance state that includes the loss of the individual's persona and social identity, which is replaced by an alien entity, usually spiritual or at least non-human" (p. 10). He also makes it clear that possessions can be found in objects as well as in living things. Through careful referencing of other studies and literature, Sersch makes the argument that possession, as defined by the above definition, is to be found in nearly every culture, and is personally believed, even today, by a large number of people (see the text for quantifications of these statements, they vary depending on the study and when the study was conducted).

Sersch brings up an important tenet that follows the narrative throughout the book—the practitioner working with a patient who believes in possession, and believes he or she to be possessed, does not have to believe in the same manner as the patient does. They only need to understand "that it is meaningful for the patient" (p. 16). At another point in the book Sersch says: " . . . a patient's belief in spirit or demonic possession does not necessitate that the therapist holds the same belief, only that the practitioner respects such a belief as valid from the worldview of the patient" (p. 22).

Sersch takes time to examine many different cultures and how those cultures view possession, what they each bring to the phenomenon, and the differing ways they approach exorcism. He also describes the primary signs of possession, first citing the Roman Catholic definition, which requires three basic criteria: 1. The possessed must speak in foreign languages or tongues, 2. They must have superhuman strength, and 3. They must know things (usually about the exorcist) that they could not possibly know. Other cultures have different criteria, some have added to these basic Roman Catholic ones. For example, in some Islamic cultures, an indication of possession includes mental and physical illness, including the hearing of voices. Sersch is also careful to differentiate between modern medical science's definitions of objective mental illness (or even physical illness) and cultural definitions. Although historically mental illness and physical illness were often considered to be caused by a possessing demon, in contemporary times exorcists are careful to rule out what the medical establishment would define as diagnosable mental illness. However, this careful scrutiny does not always find its way into an exorcism procedure, and indeed there are practitioners who still believe that the primary causal element of mental illness is possession of some external force or entity.

Sersch also speaks at length about multiple personalities (p. 11, pp. 96–101), and how individuals diagnosed with multiple personalities can be interpreted as possessed if the alternative personality is considered to be a spirit or a demon. These can be complicated distinctions and although the definition of possession, at least as defined by the Roman Catholic church, must include attributes that are not typically found in MPD (Multiple Personality Disorder) or DID (Dissociative Identity Disorder) diagnoses, they have often been included in possession research and considered in the treatment protocols (exorcism).

As mentioned earlier, a large part of Section 1 of the book deals with the history of possession throughout the world. This is a thoroughly exhaustive survey, and again quite scholarly and well-cited. Sersch points out some very interesting facts about historic possession, most notable to this reader was the view that Jesus was possessed by the Holy Spirit or the Spirit of God, in the same manner as the earlier Hebrew prophets were, an idea that was later abandoned due to its heretical nature. Enemies of Jesus

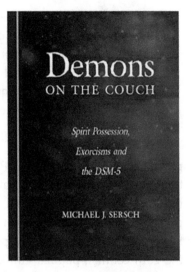

believed he was demonic (pp. 30–31). I thought it was quite interesting as well that Chinese Daoism (300 CE) used exorcism typically as a last resort to treating a person they believed was possessed. Instead, they prescribed a healthy diet, proper behavior, and spiritual practice (p. 44). That seems like good advice for treating any disease.

Sersch covers such topics as Mass Hysteria, Understanding the Witch Craze, The Standardization of Demonology, and The Faust Story. He continues with a chapter titled Modernity and Exorcism where he addresses modern ideas such as materialism: " . . . one way of understanding modernity is the shift to a material world view as a majority view, away from an enchanted world influenced by spirits and spiritual forces. This definition is especially accurate for later modernity" (p. 64). He goes on to say, "Many modern thinkers automatically dismiss all reports of possession, especially in ancient literature, as an inadequate diagnosis that can better be explained now by psychological insights" (p. 65). As mentioned earlier, even recent church-sanctioned exorcisms performed though the Roman Ritual (using the Roman Catholic church's manual of exorcising spirits) require a differential diagnosis: Is this a true possession, or a conventionally treated mental illness?

Spiro (1998) calls the scientific fallacy the belief that every phenomenon can be explained away in the mechanical–medical model. "There are dangers in reducing the experience of demonic possession to some supposedly more fundamental psychopathological condition, to neurosis, hysteria, psychosomatic disorder, and so forth" (Midelfort, 2005, p. 83). Increasingly, scholars are questioning the tendency to dismiss everything that is outside of our mechanical worldview. (p. 65)

In this chapter, much of Sersch's focus is on this materialistic paradigm of modern times. Possession simply falls outside of the boundaries of the material natural world; therefore, at best it is to be ignored, at worst ridiculed, or dismissed as some other form of psychotic mental illness. Sersch meticulously visits every area pertinent to his thesis—modernity and psychology, various faiths' exorcisms of the 20th century, popular contemporary music, and cinema. He then comments on the uprising of contemporary exorcisms, due mostly to what he calls the culture's fear of the occult (p. 81). The popularity of Blatty's book *The Exorcist* (Blatty, 1971) and the 1973 movie *The Exorcist* (Friedkin, 1972) undoubtedly adds to this fear, or maybe their popularization was due to the fear.

Sersch then comments on exorcisms gone bad, ones where the subjects were exposed to terribly abusive interventions, such as extreme restraint, or having crosses forced into their mouths (p. 82, referring to the particular case of Michael Taylor, see Ruickbie [2015]). These are primarily cases wherein a differential diagnosis was ignored or never sought, which could have concluded that the proper treatment should be more conventional (the medical treatment of a diagnosed mental illness). There have been many such "exorcisms gone bad," concentrating on physical and emotional abuse to forcibly chase out the offending demon or spirit. "Unfortunately, in many places it appears that the ancient approach of beating a person believed to be possessed in order to make the demon leave continues to be standard practice" (p. 83).

When is exorcism the right choice? As mentioned before, a patient may be a candidate for an exorcism ritual if the patient believes they are possessed and they exhibit behavior that does not fit into a more conventional diagnosis. There are many elements of the practice of exorcism that hark back to a time where formal ritual was a mainstay of human experience. Again, much of this ritualistic practice in our modern materialist culture is considered passé, old-fashioned, or superstitious. However, ritual is often considered one highly effective way to practice psychotherapy. Just sitting with a client, in a sacred space, and carefully listening, and becoming empathically tuned in to their suffering is a kind of informal ritual. Sersch suggests that finding a good exorcist to perform the ritual of exorcism is a task that requires much attention. Monsignor Andrea Gemma is interviewed by Wilkinson (2007), and in the interview Gemma says:

> So, finding someone who listens and prays is important, even psychologically. Sometimes just the fact of being listened to, or being invited into prayer and into a relationship of trust, this is a great remedy of those who are suffering. (p. 83)

Sersch continues in this chapter to explore modern exorcism covering such topics as women and possession and Catholic exorcism in the 21st century.

DIAGNOSES

The first chapter (Chapter Four) of Section 2 of the book is devoted entirely to Multiple Personality Disorder and Dissociative Identity Disorder. Here the author explains both of these disorders and their history, noting how MPD was first introduced into the Third Edition of the *DSM* in 1980 and was replaced by DID in the Fourth Edition. He cites one of the authors of the new text who was reported as saying "the reason for the change was that patients were not suffering from multiple personalities, rather they had less than one full personality" (Loftus & Ketcham, 1994). In the revised version of the *DSM-4*, variants were added to the DID diagnosis—Dissociative Disorder Not Otherwise Specified (or NOS) was a person who experienced DID but the alter was a demon or spirit. DTD, Dissociative Trance Disorder, and PTD, Possession Trance Disorder, also were added to the manual. These specifications opened up several therapeutic modalities as being acceptable methods to deal with the new designations.

By this point in the book it becomes numbingly clear that this topic is exceptionally dense and complex and covers such a multitude of topics, diversities, histories, and anecdotal experiences that it is nearly unmanageable. Again, Sersch approaches this difficulty with aplomb and confidence and thus the reader just continues to glide relatively effortlessly through it. Personally, I find the topic of exceeding importance as it touches on some very fundamental truths that I believe the culture has, since the age of materialism as mentioned before, all but obliterated from the collective consciousness. We have not yet found an effective way to define, treat, or otherwise integrate, any phenomena that do not fit neatly into the materialist paradigm. Psychology and the treatment of psychological issues are supposed to be "scientific," and that term requires adherence to laws of material cause and effect. Nowhere in psychology do we see this dichotomy of material and non-material more evident than in MPD, DID, and the now accepted variants DTD and PTD. Sersch is careful not to fall into the trap of describing these conditions in a non-scientific manner, yet he makes it clear that as practitioners we have to "act" as if what our patient is describing to us is "real." If anyone reading this has ever experienced an actual possession case and has seen, and heard, what comes from the person suffering the alleged possession, they will probably find it much easier to perceive the possession as real.

The life experience of a human being is intricately complex. The traumas a person experiences through life, both large and small, active and passive, are many and varied. If the practitioner, or patient, believes in a collective unconscious, as Carl Jung did, then the metaphysical idea is that those experiences expand beyond a post-natal life. Sersch makes great effort to explore various experiences that could be a key to a person who believes they have become possessed (see Chapter Six, Social Dynamics, p. 107). Life experiences, unconscious forces and agendas, and cultural influences that mold a particular belief system all seem to be scientific evidence for the phenomenon of possession—at least they seem to be viable explanations.

Sersch cites Bourguignon (1976) who promoted a theory that those possessed, or who claimed to be, did so as a form of role-play and that they still held a certain degree of autonomy in their execution of the possession. Sersch goes on to describe a time during his school years where he pretended to be possessed by an evil spirit. He succeeded in persuading some of his schoolmates that he was indeed possessed, but never was fully convinced himself. This story reminded me of a time in my own life when I was 13 years old and was home alone roughhousing with my dog. For whatever reason, I was inspired to take a paper grocery bag, cut holes in it for eyes, and draw on it, with crayons, a demonic face. I put the bag over my head and proceeded to growl, emit every horrible sound I could, while "attacking my dog." After a while I felt as if something was taking me over, and I became more and more aggressive, which seemed to be out of my control. The dog became terrified and ran off before any harm could be inflicted on it, leaving me in the room writhing around on the floor in my newfound demon-state. I finally got a hold of myself and pulled the bag off of my head and sat on the floor exhausted for quite some time wondering what had just happened. To this day I still wonder about it all; it was an experience I had never had, nor have had since. It does make me contemplate, and this thought is in support of some of Sersch's commentary, that my possessed state was self-induced. Possibly with the right setting, the right ritual, and the right props, anyone can call forth an evil spirit, possibly the evil shadow that resides in all of us, only waiting for the ideal moment to be known.

SUGGESTIONS FOR CLINICIANS

Section 3 begins with several interviews with exorcists. Albeit of some interest, it seemed to be a bit out of place. Chapter Ten is where the nitty-gritty begins—Contemporary Treatment. I have to admit I was a bit disappointed. I expected here an outline of an actual methodology in the treatment of a possessed patient. Sersch continues with citing the literature and further explanations of the placebo effect, how it is important the clinician be empathic to their particular worldview, and so on. The information is sound, and useful, and of course interesting. Again, I felt a strong desire to hear what Sersch himself believed, or what his conclusions were, or if he had developed some sort of methodology.

The final chapter, Conclusion, states in the first sentences that the new designations and their explanations in the *DSM-5* (American

Psychiatric Association, 2013) give practitioners the ethical, and legal, green light to treat possession. He goes on to say that any sort of exorcism, or treatment with the same intentions as exorcism, should be performed by a qualified, ethical, and experienced practitioner—a Catholic priest, a shaman, or whoever else is considered the "expert" in the particular culture or religion or belief system. Sersch goes on to say that a person should only be referred to this treatment

1. if the client believes themselves to be possessed and in need of an exorcism without coercion from the therapist or others,
2. if they have a belief system that is consistent with belief in possession states (versus clear forms of psychosis), and most importantly
3. if the ritual is performed in a safe and respectful manner, causing no harm to the person involved. (p. 139)

This is a very well-thought-out and thorough set of criteria.

In conclusion, *Demons on the Couch* is an excellent book. It is very well-written, incredibly well-cited and referenced, and contains just about everything a reader would want to know about possession through the lens of a practicing psychologist or psychotherapist. It would be accessible and useful for anyone without ruffling any feathers regarding belief or superstition, as Sersch makes it evident, as he states at the beginning of the book, that his intention is not to prove or disprove the reality of possession, demons, spirits, or the like.

As I stated earlier, I would have liked to have heard a bit more from Sersch himself regarding his own experiences. He tantalizes us a bit with a few anecdotal references, but it is not enough in my opinion. I also would have liked to have seen some sort of discussion about working with a possessed client without having to perform an actual exorcism, but I can understand if Sersch was intentionally avoiding that possible pit.

REFERENCES

American Psychiatric Association. (1994). *The diagnostic and statistical manual of mental disorders* (4th ed.). American Psychiatric Association.

American Psychiatric Association. (2013). *The diagnostic and statistical manual of mental disorders* (5th ed.). American Psychiatric Association.

Blatty, W. P. (1971). *The exorcist.* Harper & Row.

Bourguignon, E. (1976). *Possession.* Chandler and Sharp.

Friedkin, W. (Director). (1973). *The exorcist* [Motion Picture]. Warner Bros. Blatty, W. P., & Marshall, N. (Producers).

Loftus, E., & Ketcham, K. (1994). *The myth of repressed memory: False memories and allegations of sexual abuse.* St. Martin's Griffin.

Lund, H. (2014). Spirituality, religion and faith in psychotherapy practice. In M. Natanson (Ed.), *Phenomenology and the social sciences.* Northwestern University Press.

Midelfort, H. C. E. (2005). *Exorcism and enlightenment: Johann Joseph Gassner and the demons of eighteenth-century Germany.* Yale University Press.

Ruickbie, L. (2015). "Taylor, Michael." In J. P. Laycock (Ed.), *Spirit possession around the world: Possession, communion, and demon expulsion across cultures* (pp. 337–339). ABC-CLIO.

Spiro, H. (1998). *The power of hope: A doctor's perspective.* Yale University Press.

Wilkinson, T. (2007). *The Vatican's exorcists: Driving out the devil in the 21st century.* Warner.

Journal of Scientific Exploration, Vol. 34, No. 2, pp. 392–394, 2020   0892-3310/20

BOOK REVIEW

Enchanted Ground: The Spirit Room of Jonathan Koons
by Sharon Hatfield. Swallow Press, 2018. 342+16 pp. $23.16
(hardcover). ISBN: 978-0804012089.

REVIEWED BY TOM RUFFLES

tom.ruffles@googlemail.com

https://doi.org/10.31275/2020/1791

The mediumship of Jonathan Koons is much less well-known than that of the Fox sisters, though during the 1850s his Spirit Room at Athens, Ohio, became a desirable destination for Spiritualists keen to seek contact with the spirit world. There they experienced a wide variety of phenomena that would shortly become staples of the movement. Local resident and nonfiction author Sharon Hatfield has made a painstaking examination of Koons's life and career based on the available evidence, setting it in the context of the developing religion of Spiritualism.

Koons was born in Pennsylvania in 1811. He moved to Mount Nebo—actually a large hill in Athens County, southeastern Ohio, in the foothills of the Appalachians—in 1835, where he farmed and became patriarch to a sizeable family. In his youth he had had a mystical experience involving a 'visit to the realm of light' which led him to believe he had met an angel. Originally Presbyterian, he abandoned the religion as he found its Calvinism uncongenial, becoming as he put it an 'infidel.' He lacked a formal education but read widely.

In early 1852 he encountered reports of the Fox sisters' mediumship and decided to investigate the subject. He attended séances and, initially skeptical, was told he himself had a gift. Trying it, he found a new religious faith. He also discovered that other members of his family, particularly his eldest son Nahum, had mediumistic abilities. After a while, they were told by Spirit to build a dedicated room to hold about 20–25 people, indicating that séances were to include more than his immediate friends and family. Koons duly built a log cabin

on the farm to the given specifications
and furnished it, including with musical
instruments, as instructed. The stage was
then set for public séances.

These soon attracted the attention of
both truth seekers and the curious from
a widening geographical area. It was an
isolated spot and not easy to reach, despite
which many made the journey; the book's
title comes from the account of a visitor in
1855 who described traveling 'towards the
enchanted ground.' Koons did not charge
admission, and while those who stayed
were expected to pay something toward bed and board, this generated
very little income, and it seems likely that the family subsidized the
operation of the Spirit Room.

Séances were conducted in total darkness and were lively. Koons
played his fiddle, and soon there would be an accompaniment from
a spirit band that moved around the crowded room playing loudly.
A trumpet allowed spirits to speak directly to the sitters. Spirit
hands, feeling cold to the touch, appeared, either self-illuminating
or illuminated by sheets of phosphorous-covered paper. Lengthy
messages, written very quickly, were conveyed to sitters.

As a result, Koons gained a prominent and generally respected
position in the movement, with his activities widely reported in the
Spiritualist press, though he had detractors as well as enthusiastic
champions. He was a prolifically polemical writer, further raising his
profile. The Spirit Room was in existence for a relatively short period
and had closed by November 1855, though the family continued to hold
séances elsewhere for a time. In 1858 Koons and his family moved to
Illinois, where he died in 1893.

In foregrounding Koons's career in Spiritualism, Hatfield has
provided a fascinating snapshot of the movement in its infancy. She
does not reach a conclusion on his claim of contact with the spirit
world, focusing instead on ritual and belief and the transformative
effect of Koons's séances, but the issue of fraud always hovers in the
background despite the many testimonies from eyewitnesses who

claimed fraud had been precluded by the physical difficulties such a confined space presented.

Other attendees were less impressed, noting Koons's resistance to the imposition of thorough controls and the darkness within which the family operated. It was not unusual for mediums producing seemingly impeccable phenomena to be unmasked as cheats, and Koons's reputation was damaged by an event in Cleveland in 1856: The family was holding a séance at the home of the editor of the *Spiritual Universe* when a suddenly struck match showed his daughter Quintilla on her feet.

Despite his critics' negative views, Hatfield, following Brandon Hodge's (2015) similar conclusion, demonstrates that Koons was a significant force in early Spiritualism rather than the peripheral figure he had previously been considered. This is a welcome corrective because, notwithstanding his inclusion in Emma Hardinge Britten's *Modern American Spiritualism*, for someone who played a seminal role in formulating the repertoire of the dark séance and was an influence on later mediums such as the Davenport brothers, he has been largely neglected in more recent scholarship. Even Bret E. Carroll's *Spiritualism in Antebellum America*, a detailed examination of the period during which the Spirit Room flourished, devotes only two pages to Koons.

To some extent this neglect may have been because commentators shared Sir Arthur Conan Doyle's prejudice when he dismissed Koons as "a case of true physical mediumship of a crude quality, as might be expected where a rude uncultured farmer was the physical centre of it" (1926). Hatfield has redressed the balance with this valuable addition to the literature, and she amply justifies her claim, referring to the Spiritualist movement, that Koons was "one of its most charismatic figures—a backwoods seer whose legacy would rival even that of the famous Fox sisters for a place in its history."

REFERENCES

Britten, E. H. (1870). *Modern American Spiritualism: A twenty years' record of the communion between Earth and the world of spirits.* Self-published.

Carroll, B. E. (1997). *Spiritualism in antebellum America.* Indiana University Press.

Conan Doyle, A. (1926). *The history of spiritualism.* Cassell & Co.

Hodge, B. (2015). Ghosts in the machines. 'For want of a nail.' *Paranormal Review,* 75, 34–35; 76, 36–37.

39th Annual SSE Conference

Combined Conference with the Parapsychological Association

hosted by the Rhine Research Center

June 2021

Hilton Durham Hotel, Durham, North Carolina

near Duke University

2021 VISION: LOOKING FORWARD
Experience, Experiments, Beyond

Forthcoming detailed information:

https://societyforscientificexploration.org/conferences/2021

SSE ASPIRING EXPLORERS PROGRAM

The SSE has established Aspiring Explorers Awards for meritorious student research projects judged to be the most original and well-executed submissions in subject areas of interest to the SSE. A committee is in place to review all entries and determine the winners, who will receive awards of $500 each. One award winner will have the opportunity to present a talk describing the project at the SSE Annual Meeting, for which the Society will cover his/her registration fee. The other award winner will have the opportunity to present a talk describing their project at the SSE Euro Meeting, for which the Society will cover her/his registration fee. Submissions must be made per the guidelines and deadline as stated on the SSE website "Call for Papers" for the conference you are considering attending in order to be eligible for that year's prize for that conference.

If your paper is selected for the Aspiring Explorer Award, you will be either invited to present your talk at the meeting or able to submit your paper as a poster session. We are very excited about the recent poster sessions at annual SSE meeting, so please let your fellow student colleagues and professors know about this. https://societyforscientificexploration.org/conferences/2021

In addition, the SSE is also offering a 50% discount on future meeting registrations for any student member who brings one student friend to our conferences (one discount per student). We are eager to see student clubs or SSE discussion groups established at various academic institutions or in local communities. Contact us at education@scientificexploration.org to start your own group!

C. M. Chantal Toporow, Ph.D., SSE Education Officer
education@scientificexploration.org

FUNDAÇÃO
Bial
Institution of public utility

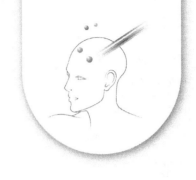

Funding for Scientific Research
2020/2021

With the aim of encouraging the research into healthy human being's physical and mental processes, namely in fields still largely unexplored but which warrant further scientific analysis, BIAL Foundation opens now a Grants programme for Scientific Research with the following characteristics:

1. Scope and purpose - Only the fields of Psychophysiology and Parapsychology shall be covered by this programme. The goals to be met by the applicants shall be set out by the Research Project under application.

2. Addressees - All scientific researchers will be admitted as applicants, either individually or in groups, except those working for BIAL Foundation or for any of the companies belonging to BIAL Group. The current Grant Holders of BIAL Foundation can also be admitted as applicants; however, they shall only benefit from new grants under this programme after the successful completion of the work comprised in the scope of previous awarded grants.

3. Duration and commencement - The total duration of the grants shall not exceed 3 years and shall commence between 1st of January and 31st of October 2021.

4. Total amount and payment's periodicity - The approved applications shall benefit from grants in total amounts comprised between €5,000 and €50,000. The specific amount shall be fixed at BIAL Foundation's sole discretion in accordance with the needs of the Research Project under application.
The amount awarded to each Research Project shall be understood as a maximum amount, which shall be paid by BIAL Foundation upon verification of the documents of expenses submitted.

The payments shall be made annually or bi-annually. This periodicity shall be defined in accordance with the schedule of the Research Project.

5. Applications - Applications should be submitted in English no later than 31st of August 2020, in accordance with the Regulation of Grants for Scientific Research of BIAL Foundation, via specific online application form available at www.fundacaobial.com. Applications of projects from Clinical or Experimental Models of Human Disease and Therapy shall not be accepted.

6. Assessment of applications and disclosure of results - Applications shall be assessed by the Scientific Board of BIAL Foundation. The decision shall be disclosed, by notice to the applicants, within 4 (four) months from the final deadline for submission of applications mentioned in the preceding section 5.

7. Applicable Regulation – The submission of an application implies the full acceptance by the applicant of the terms and conditions set out in this announcement and in the Regulation of Grants for Scientific Research of BIAL Foundation, which governs the present programme.

BIAL Foundation reserves the right to refuse the application of former Grant Holders who have repeatedly violated their legal and contractual obligations with BIAL Foundation.

The Regulation of Grants for Scientific Research of the BIAL Foundation is available at:

A Av. da Siderurgia Nacional
4745-457 Coronado (S. Romão e S. Mamede) • Portugal
Tel. + 351 22 986 6100 • Fax + 351 22 986 6199
www.fundacaobial.com • fundacao@bial.com

Society for Scientific Exploration

Index of Previous Articles in the Journal of Scientific Exploration

SOCIETY FOR
SCIENTIFIC EXPLORATION

GIFT *JSE* ISSUES & SSE MEMBERSHIPS, BACK ISSUES

Single Issue: A single copy of this issue can be purchased.
                    For back issues from 2010 to present in print, contact
                    membership@scientificexploration.org
Price: $20.00 EACH

Online Access: **All issues and articles from Volume 1 to present can be**
                    read for free at https://www.scientificexploration.org/journal-library
Subscription: To send a gift subscription, fill out the form below.
Price: $85 (online)/$145(print & online) for an individual
                    $165/$225 (online only/print) for library/institution/business

Gift Recipient Name _____
Gift Recipient Address _____

_____

Email Address of Recipient for publications_____
☐   I wish to purchase a single copy of the current JSE issue.
☐   I wish to remain anonymous to the gift recipient.
☐   I wish to give a gift subscription to a library chosen by SSE.
☐   I wish to give myself a subscription.
☐   I wish to join SSE as an Associate Member for $85/yr or $145/yr
    (print), and receive this
    quarterly journal (plus the EdgeScience magazine and The Explorer
    newsletter).

Your Name _____
Your Address _____

_____

Send this form to: Journal of Scientific Exploration
                     Society for Scientific Exploration
                     P. O. Box 8012
                     Princeton, NJ 08543-8012

                     Membership@ScientificExploration.org
                     Phone (1) (609) 349-8059

For more information about the Journal and the Society, go to
https://www.scientificexploration.org

**SOCIETY FOR
SCIENTIFIC EXPLORATION**

JOIN THE SOCIETY AS A MEMBER

Join online at https://www.scientificexploration.org/join

The Society for Scientific Exploration has four member types:

Associate Member ($85/year with online Journal; $145 includes print Journal): Anyone who supports the goals of the Society is welcome. No application material is required.

Student Member ($40/year includes online Journal; $100/year includes print Journal): Send proof of enrollment in an accredited educational institution.

Full Member ($125/year for online Journal; $185/year includes Print Journal): Full Members may vote in SSE elections, hold office in SSE, and present papers at annual conferences. Full Members are scientists or other scholars who have an established reputation in their field of study. Most Full Members have: a) doctoral degree or equivalent; b) appointment at a university, college, or other research institution; and c) a record of publication in the traditional scholarly literature. Application material required: 1) Your curriculum vitae; 2) Bibliography of your publications; 2) Supporting statement by a Full SSE member; or the name of Full SSE member we may contact for support of your application. Send application materials to SSE Secretary Mark Urban-Lurain, scientificexplorationsecretary@gmail.com

Emeritus Member ($85/year with online Journal; $145/year includes print Journal): Full Members who are now retired persons may apply for Emeritus Status. Please send birth year, retirement year, and institution/company retired from.

All SSE members receive: online quarterly *Journal of Scientific Exploration* (JSE), *EdgeScience* online magazine, *The Explorer* online newsletter, notices of conferences, access to SSE online services, searchable recent *Journal* articles from 2008–current on the JSE site by member password, and for all 25 years of previous articles on the SSE site. For additional new benefits, see World Institute for Scientific Exploration, instituteforscientificexploration.org

---

Your Name _____

_____

Email_____

Phone_____ Fax_____

Payment of _____ enclosed, *or*

    Charge My   VISA ☐   Mastercard ☐

    Card Number _____ Expiration_____

    Send this form to:
        Society for Scientific Exploration
        P. O. Box 8012
        Princeton, NJ 08543-8012
        Membership@ScientificExploration.org
        Phone (1) (609) 349-8059

For more information about the Journal *and the Society, go to*
https://www.scientificexploration.org

JOURNAL OF SCIENTIFIC EXPLORATION
A Publication of the Society for Scientific Exploration

Instructions for Authors (Revised October 2018)

JSE is an open access journal, with a CC-BY-NC license. There are no submission fees, no reader fees, no author page charges, and no open access fees. Please submit all manuscripts at http://journalofscientificexploration.org/index.php/jse/login (please note that "www" is NOT used in this address). This website provides directions for author registration and online submission of manuscripts. Full Author Instructions are posted on the Society for Scientific Exploration's website at https://www.scientificexploration.org/documents/instructions_for_authors.pdf for submission of items for publication in the *Journal of Scientific Exploration* (including "Writing the Empirical Journal Article." Before you submit a paper, please familiarize yourself with the *Journal* by reading JSE articles. Back issues can be read at https://www.scientificexploration.org/journal-library, scroll down to the Open Access issues. Electronic files of text, tables, and figures at resolution of a minimum of 300 dpi (TIF or PDF preferred) will be required for online submission. You will also need to attest to a statement online that the article has not been previously published and is not submitted elsewhere. The journal uses the *APA Publication Manual 7th edition* as its style guide.

JSE Managing Editor: Journal@ScientificExploration.org

AIMS AND SCOPE: The *Journal of Scientific Exploration* publishes material consistent with the Society's mission: to provide a professional forum for critical discussion of topics that are for various reasons ignored or studied inadequately within mainstream science, and to promote improved understanding of social and intellectual factors that limit the scope of scientific inquiry. Topics of interest cover a wide spectrum, ranging from apparent anomalies in well-established disciplines to rogue phenomena that seem to belong to no established discipline, as well as philosophical issues about the connections among disciplines. The *Journal* publishes research articles, review articles, essays, book reviews, and letters or commentaries pertaining to previously published material.

REFEREEING: Manuscripts will be sent to one or more referees at the discretion of the Editor-in-Chief. Reviewers are given the option of providing an anonymous report or a signed report.

In established disciplines, concordance with accepted disciplinary paradigms is the chief guide in evaluating material for scholarly publication. On many of the matters of interest to the Society for Scientific Exploration, however, consensus does not prevail. Therefore the *Journal of Scientific Exploration* necessarily publishes claimed observations and proffered explanations that will seem more speculative or less plausible than those appearing in some mainstream disciplinary journals. Nevertheless, those observations and explanations must conform to rigorous standards of observational techniques and logical argument.

If publication is deemed warranted but there remain points of disagreement between authors and referee(s), the reviewer(s) may be given the option of having their opinion(s) published along with the article, subject to the Editor-in-Chief's judgment as to length, wording, and the like. The publication of such critical reviews is intended to encourage debate and discussion of controversial issues, since such debate and discussion offer the only path toward eventual resolution and consensus.

LETTERS TO THE EDITOR intended for publication should be clearly identified as such. They should be directed strictly to the point at issue, as concisely as possible, and will be published, possibly in edited form, at the discretion of the Editor-in-Chief.

PROOFS AND AUTHOR COPIES: Authors will receive copyedited, typeset page proofs for review. Print copies of the published Journal will be sent to all named authors if requested.

COPYRIGHT: Authors retain copyright to their writings. However, when an article has been submitted to the *Journal of Scientific Exploration* for consideration, the *Journal* holds first serial (periodical) publication rights. Additionally, after acceptance and publication, the Society has the right to post the article on the Internet and to make it available via electronic as well as print subscription. The material must not appear anywhere else (including on an Internet website) until it has been published by the *Journal* (or rejected for publication). After publication in the *Journal*, authors may use the material as they wish but should make appropriate reference to the prior publication in the *Journal*. For example: "Reprinted from [or From] "[title of article]", *Journal of Scientific Exploration*, vol. [xx], no. [xx], pp. [xx], published by the Society for Scientific Exploration, http://www.scientificexploration.org."

DISCLAIMER: While every effort is made by the Publisher, Editors, and Editorial Board to see that no inaccurate or misleading data, opinion, or statement appears in this *Journal*, they wish to point out that the data and opinions appearing in the articles and announcements herein are the sole responsibility of the contributor concerned. The Publisher, Editors, Editorial Board, and their respective employees, officers, and agents accept no responsibility or liability for the consequences of any such inaccurate or misleading data, opinion, or statement.